Frederick Jackson Turner, Dept. of Agriculture New Sout
Wales

A Census of the Grasses of New South Wales

Together With a Popular Description of Each Species

Frederick Jackson Turner, Dept. of Agriculture New Sout Wales

A Census of the Grasses of New South Wales
Together With a Popular Description of Each Species

ISBN/EAN: 9783744692083

Printed in Europe, USA, Canada, Australia, Japan

Cover: Foto ©berggeist007 / pixelio.de

More available books at **www.hansebooks.com**

OF THE

GRASSES OF NEW SOUTH WALES

TOGETHER WITH A

POPULAR DESCRIPTION OF EACH SPECIES;

BY

FRED. TURNER, F.R.H.S.,

BOTANIST TO THE DEPARTMENT OF AGRICULTURE, NEW SOUTH WALES.

ISSUED BY DIRECTION OF

THE MINISTER FOR MINES AND AGRICULTURE
(HON. SYDNEY SMITH).

H. C. L. ANDERSON, DIRECTOR.

SYDNEY: CHARLES POTTER, GOVERNMENT PRINTER.

1890.
[1s.]

11b 105—90

Astrebla triticoides, F.v.M. "Mitchell Grass."

PREFACE.

To popularise any branch of natural science so as to make it acceptable to the general public it becomes necessary to divest it of as much technical language as possible, without, however, interfering with the scientific name of anything. Some scientific nomenclature must be given to make any object universally understood. It has been often remarked that no books are more ridiculed by the general public than those dealing with science. Keeping these thoughts in mind the author ventures to write a popular description of all the grasses indigenous to New South Wales, so that the information given may be as easily understood in the small settler's home as in the more pretentious dwelling of the pastoralist. In 1879 the writer compiled his first list of 100 Australian grasses, together with a popular description of each species. This led to so much interest being taken in them, both by pastoralists and farmers, that the writer has, since that time, named thousands of specimens sent to him for identification, and given much information as regards their value for forage, hay, &c. It is thought that the present time is most opportune for issuing a full description of all our grasses, so that the latest information regarding their merits may be placed before the public. After the abundant rainfall we have had, the natural grasses will (unless in places where they have been too closely fed down, which, however, is their general condition) produce an abundance of seed, and there should be no difficulty in collecting any quantity in the railway reserves, or in specially reserved areas, for dissemination throughout the country. The time of the year at which each grass ripens its seeds is noted, and also the district or locality where it grows, and whether it is an annual or perennial, and as far as is known its common name is given in the descriptive matter. This will help the collector to identify some of the species. It is generally admitted by all observant persons interested in agricultural and pastoral pursuits that something in the way of systematic cultivation or conservation must be undertaken ere long to save some of our most valuable grasses from extinction. There is no gainsaying the fact that the overstocking of runs has had a most serious effect upon the pastures in this Colony. It not only has destroyed many valuable salinous and other forage plants and grasses, but many useless weeds have sprung up in their places which stock will not touch. Some of these weeds have proved such a pest to the country that legislation has been directed towards their extermination. If this Colony intends to hold her own against the world in the production of high-class wool, also in the matter of the frozen meat export trade, it becomes of vital importance to the population that more attention should be paid to our pastures than has hitherto been the case. That we have far more valuable native grasses in this country than any yet introduced I have abundant proof, having experimented on upwards of 100 species of European, Asiatic, African, American, and Australian grasses, to test their true qualities by comparison. In these trials the bulk of the Australian grasses yielded more forage and withstood the drought much better than did the exotics. If all the money that has been sent out of the Colony to import exotic grass seeds (which often prove a failure in this land of drought and flood) had been expended on the cultivation, or even systematic conservation, of our native grasses and other herbage, we should not periodically hear of thousands of cattle and sheep dying of starvation during every recurring drought.

As far as is at present known there are 197 species and thirty-three well-defined varieties of grasses indigenous in New South Wales. Out of this number there are sixteen species which are dreaded by pastoralists on account of their long seed awns or sharp pointed leaves; not a very formidable array it must be admitted, still of sufficient importance to make their position felt and somewhat dreaded in the country.

The Honorable S. Smith, Minister for Mines and Agriculture, has given instructions for the preparation of an illustrated monograph of the grasses of New South Wales, which will appear in the *Agricultural Gazette*. Those persons reading this little work can assist in this most useful undertaking by forwarding good specimens of grasses, with notes, and locality where collected, to the Director of Agriculture, Phillip-street, Sydney, to whom also any botanical specimens, which should be in flower and fruit, may be sent for identification.

The Department has now in hand the preparation of a list of all the introduced weeds, with instructions given as to the best means of exterminating them. And at no distant date a list with descriptive matter will be published of all the supposed poisonous plants, indigenous in New South Wales. Pastoralists, farmers, and others, can materially assist the Department to carry out this work to a successful issue.

At the end of this little work will be found an appendix, containing a description of the exotic grasses, that have become acclimatised, and now are apparently wild in the Colony.

A few notes on collecting grasses :—A specimen of grass, to be perfect, should have roots, stems, leaves, flowers, and fruit (seeds). Fragmentary specimens, such as leaves without flowers, or flowers without leaves, are of little or no use. When a specimen (average size) is found it should be carefully pulled, or dug up, so as to preserve as many roots and leaves as possible ; after shaking the soil from the roots lay each specimen between sheets of brown paper, or ordinary newspaper will do ; then lay the sheets of paper containing the specimens between two strong boards a little longer and wider than the specimens, and subject them to a slight pressure at first; every alternate day the specimens should be changed into perfectly dry paper, and subjected to increased pressure until they are quite dry. Under ordinary circumstances the operation will not take longer than eight or nine days. The exact amount of pressure to be given, however, will depend on the consistence of the grass. It must be borne in mind that too much pressure crushes the delicate parts, and too little allows them to shrivel. Stones. or any pieces of old iron, will answer as weights for pressing the specimens.

F. TURNER, Botanist,
Department of Agriculture.

A Census of the Grasses of New South Wales.

AGROPYRUM PECTINATUM, Beauv.
"Maneroo wheat grass."

A SLENDER grass, when found on poor soils, but on rich ground it will grow 18 inches or 2 feet high, and yields a fair amount of good herbage. When in a young state it is much relished by stock. As it becomes old, however, it is rather harsh; then stock will seldom or never touch it whilst other feed is plentiful. If cut, when it shows its flower spikes, it ought to make excellent hay. When left unmolested, it will produce a fair amount of seed, which ripens during November and December. Habitat : Maneroo Plains and the more southern portions of the Colony.

AGROPYRUM SCABRUM, Beauv.
"Wheat grass."

A most variable grass as regards height. On poor soils it will rarely exceed 1 foot, but on rich land it often grows 3 feet high. During the winter, and early spring months, ere many of our native grasses begin to grow, it yields a rich succulent herbage, which stock of all kinds are fond of. When it becomes old, however, it is rather harsh. If cut when it shows its flower spikes, it makes excellent hay. 1 have had this grass under cultivation, and it was much superior to that ordinarily seen in pastures, both in bulk and quality. I can recommend its cultivation either for early spring feed, or for making into hay. When the seeds are ripe, its seed awns are often troublesome, causing irritation to lambs, by getting into their eyes and wool. When left unmolested for a time it will produce a fair amount of seed, which ripens during August, September, and October. There is a variety of this grass with shorter seed awns. With this exception, however, its qualities are much the same. Habitat : generally all over the Colony.

AGROPYRUM VELUTINUM, Nees.
"Velvet-leaved wheat grass."

A dwarf species, with soft hairy leaves, rarely exceeding 1 foot in height. When other feed is plentiful, it is not much relished by stock. It generally grows on land liable to inundation in the southern parts of the Colony. It produces a fair amount of seed, which ripens during October, November, and December.

AGROSTIS MUELLERI, Benth.
"Bent grass."

An alpine (densely tufted) grass, rarely exceeding 6 inches high. During the summer months it yields a fair amount of succulent herbage, of which sheep are particularly fond. If not cropped too closely down it will produce a fair amount of seed, which ripens during December and January. Perennial, Habitat : Southern mountains, New South Wales.

B

AGROSTIS SCABRA, Willd.

"Bent grass."

A slender tufted grass, growing from 6 inches to 1½ feet high. On the high table-lands, and in the colder parts of the Colony, it yields a fair amount of rich herbage, which sheep are fond of. The seeds ripen in November, December, and January.

AGROSTIS VENUSTA, Trin.

"Bent grass."

A grass of somewhat similar habit to the last, but it is found more abundantly in the low land pastures in the southern parts of the Colony. Seeds in November and December.

AIRA CARYOPHYLLEA, Linn.

"Silver-haired grass."

A slender elegant tufted annual, rarely exceeding 6 inches in height. It is generally found in the southern portions of the Colony, although the Rev. Dr. Woolls collected specimens near Mudgee a few years ago. It is not of much value as a pasture grass, but it is worth cultivating in gardens for its very ornamental appearance. The seeds ripen in September and October.

ALOPECURUS GENICULATUS, Linn.

"Knee-jointed fox-tail grass."

A weak-growing annual species, with procumbent stems at the base, and sometimes ascending to 1½ feet high. It is found very plentifully around shallow pools of water in the western districts during the winter and early spring months, and produces a rich succulent herbage, which is much relished by stock of all kinds. The seeds ripen in September and October.

AMPHIBROMUS NEESII, Steud.

"Marsh brome grass."

A succulent perennial grass, growing from 3 to 5 feet high, and is generally found in and around shallow pools of water, both in the coastal and western districts. Its rich herbage is greedily eaten by stock of all kinds. On low marshy lands, where few other kinds of grasses would thrive, this species would pay to cultivate for making into ensilage, or keeping it in reserve for a hot dry summer, when other feed is scarce. Where cattle are prevented from eating this grass for a time it will produce a great amount of seed, so there should be no difficulty in disseminating it throughout the Colony. The seeds ripen during September, October, and November.

AMPHIPOGON STRICTUS, R. Br.

"Bearded-head grass."

An erect, slender, perennial growing species, from 1 foot to 2 feet high. There are two or three varieties of this grass, none of them, however, affording very much herbage. It is a good grass for standing a long drought, and at such times, when herbage is scarce, it proves of some value in the pastures. The seeds ripen in November, December, and January. Habitat: generally all over the Colony.

ANDROPOGON AFFINIS, R. Br.
"Blue grass."

A perennial species, growing from 1 foot to 2 feet high. It is a valuable pasture grass, and grows nearly all the year round in the coastal districts. It yields a fair amount of rich herbage, much relished by all herbivora. This grass stands the drought well, and will bear close feeding. It will also make good hay. The seeds are produced very freely, and they ripen during the summer months.

ANDROPOGON BOMBYCINUS, R. Br.
"Silky heads."

An erect, rigid, perennial grass, growing from 1 foot to 3 feet high. Stock are remarkably fond of it when young, but when it gets old it is somewhat harsh, and they will leave it for more tender herbage. When it is brought under cultivation, however, it is a most prolific grass, and it loses that harshness even when it gets old, that characterises it when grown on uncultivated land. It will make capital hay. It is one of those grasses, the roots of which penetrate deeply into the soil, and it resists the drought to a marked degree. I have seen this grass bearing seed in the western districts when many other species had withered off the land through drought. The seeds ripen in November, December, and January. Habitat: generally over the western interior.

ANDROPOGON ERIANTHOIDES, F.v.M.
"Satin heads."

An erect glaucous perennial grass, growing from 2 to 3 feet high. A very superior pasture grass, which is in danger of becoming extinct. Stock of all descriptions thrive well upon it, and often crop it down so close, that it has little chance to recuperate, or perfect seeds for its natural reproduction. It is one of those grasses that root deeply into the soil, which enables it to withstand a long drought. I have had this species also under cultivation, and it yielded a great amount of rich succulent herbage. It makes splendid hay, and I can highly recommend its cultivation to anyone. When left unmolested for a time, it will produce a great amount of seed, which ripens during November and December. Habitat: Western Districts.

ANDROPOGON INTERMEDIUS, R. Br.
"A Blue grass."

An erect perennial grass, growing from 2 to 3 feet high. It produces a quantity of coarse herbage, which, however, is readily eaten by stock of all descriptions. It would also make capital hay, if cut just as it is showing its flowers. This grass is generally found on land that is liable to periodical inundations, and on the borders of rivers where it grows into large tussocks. It can be recommended for binding the banks of dams, and any loose earth. The seeds ripen during November, December, and January. Habitat: North-western Districts.

ANDROPOGON LACHNATHERUS, Benth.
"Hairy-headed blue grass."

A rather slender erect perennial grass, growing about 2 feet high. During the early part of the summer, it affords a fair amount of herbage much relished by stock. When it sends up its wiry flower-stems, however, cattle seldom or never touch it. This grass will make good hay if cut young. It is found on low-lying rich soils near rivers and lagoons in the north-eastern portion of the Colony. The seeds ripen during November and December.

ANDROPOGON PERTUSUS, Willd.

"Pitted blue grass."

A slender erect perennial grass, growing from 1 foot to 3 feet high. It stands the drought well, and it will grow throughout the winter months in the coastal districts where it is not too cold. It is an excellent pasture grass, and yields a fair amount of herbage which cattle and sheep are very fond of. It is much improved by cultivation, and if cut when it shows its flowers, makes good hay. The seeds ripen in October and November. Found over a greater portion of the Colony.

ANDROPOGON REFRACTUS, R. Br.

"A Kangaroo grass."

The base of the stems and roots of this species like that of several others of the genus, are highly aromatic. During the summer months, it makes a great amount of herbage, which is relished by all pasture animals. It is a most productive grass when under cultivation, and if cut when it shows its flower stems, makes excellent hay, with a slightly aromatic perfume. It is not particular as to soil or situation, for it may often be found growing on dry stony ridges, and on rich alluvial soils. On the latter kind of land, however, it produces a better herbage, and it will grow from 3 to 4 feet high, It is perennial, and produces plenty of seed, which ripens during November and December. Found over a greater portion of the Colony.

ANDROPOGON SERICEUS, R. Br.

"The Blue grass" of the Colonists.

An erect perennial grass, growing from 1 foot to 2½ feet high. It is generally found growing on rich soils, over a greater portion of the Colony. It is a most productive grass, and during the summer months yields a rich succulent herbage, much relished by all herbivora. Having had this grass under cultivation, I can highly recommend it for permanent pasture or hay-making—for the latter particularly so. It perfects a great amount of seed, which germinates readily under ordinary conditions, and owing to these circumstances we may account for such a valuable forage plant still being fairly plentiful in some situations. The seeds ripen during October and November.

ANISOPOGON AVENACEUS, R. Br.

By some persons known as the "Oat grass."

A tall, glabrous grass, growing from 2 to 3 feet high, and branching at the base only. It produces a fair amount of leafy bottom forage, and when young, catttle eat it down readily. It yields a fair-sized grain, which might be improved by cultivation. Several kinds of birds eat the grain, as also do fowls. Seeds during November, December, and January. This grass has not a wide geographical range—as far as I am aware it has not been found west of the Blue Mountains.

ANTHISTIRIA AVENACEA, F.v.M.

"Tall oat, or Kangaroo grass."

The stems of this perennial grass rise from a woolly, thick base, to a height of 3 or 4, and sometimes 5 feet. It is found growing in tussocks, only on the richest of soils in the interior, and is fairly plentiful in some situations. This grass produces a large quantity of good leafy feed at the base, which cattle are remarkably fond of. In a young state it is very nutritious, but when it gets old the flower stems become hard and cane-like; then cattle seldom touch it. Its roots penetrate deeply into the soil, which enables the plant to withstand a long spell of dry weather with impunity. This grass might be profitably cultivated for ensilage if it were cut before the flower stalks became so hard and cane-like. Unlike some other kangaroo grasses, it possesses the advantage of being a prolific seeder. The grains are like small oats, and ripen during November and December.

ANTHISTIRIA CILIATA, Linn.

The common " Kangaroo grass."

A perennial upright growing grass, often over 3 feet in height, when found on rich soils. It enjoys a wide reputation of being one of the finest and most useful of the indigenous grasses on the eastern side of the dividing range, stock of all descriptions being remarkably fond of it. The roots are strong, and penetrate the soil to a great depth, which enables the plant to remain green during the greater part of the summer. In the autumn the foliage turns brown, when, however, its nutritive qualities are said to be at the highest. If cut as the flower stems appear it can be made into excellent hay. Although this grass throws up a number of flowering stems, still it perfects very little seed, and the most reliable way to propagate it is by division of the roots. This may seem to be a tedious process, but it would soon pay for the outlay by the immense yield it would give in a very short time. Found all over the Colony. In the coastal districts it is plentiful, but in the interior it is more sparingly distributed.

Baron von Mueller and L. Rummel give the following chemical analysis of this grass during its spring growth:—Albumen, 2·05; gluten, 4·67; starch, 0·69; gum, 1·67; sugar, 3·06 per cent.

ANTHISTIRIA MEMBRANACEA, Lindl.

"Landsborough grass."

An annual grass, sometimes growing in small tufts. In favourable seasons, however, the stems are long and weak, forming an entangled mass over a foot deep. It is a quick-growing summer species, and is particularly adapted for hot, dry regions. Towards autumn it gets exceedingly dry and brittle, but stock are so fond of it that they often lick the broken parts from the ground. It is a most nutritious grass, and is worthy of extensive culti-vation. I have grown this grass from seed and had a good crop in less than three months. It makes capital hay. It produces an abundance of seed, which ripens in November and December. Found nearly all over the western interior.

APLUDA MUTICA, Linn.

A species with creeping or climbing stems, often several feet long, with erect branching flowering shoots and long leaves; not of much value for forage. A rare species.

ARISTIDA ARENARIA, Gaud.

" Three-awned spear grass."

A perennial species found on rich loamy soils in the interior, but only sparingly eaten by stock when young. When it becomes old and goes to seed it is of a dry, wiry, nature, and the seed awns adhere to the palates and throats of the animals eating them. The awns and barbed seeds not only injure the wool, but often penetrate the skins, and reach the vital parts of the animals. This grass seeds in October and November.

ARISTIDA BEHRIANA, F. v. M.
"Three-awned spear grass."

A perennial species, usually growing about 1½ feet high; of a similar character to the last species; found in the western interior. Seeds in November and December.

ARISTIDA CALYCINA, R. Br.
"Three-awned spear grass."

A perennial species, growing from 1 foot to 2½ feet high. Found principally on the sand-hills of the interior. It is only eaten by stock during a scarcity of other herbage. The seeds, with their adherent awns, are injurious to wool, and troublesome to the eyes of sheep. It is a grass, however, that improves under cultivation, for the plant loses much of the rigidity common to uncultivated ones. The seeds ripen during November and December.

ARISTIDA DEPRESSA, Retz.
"Three-awned spear grass."

A perennial species, generally found growing on rich soils, but of too harsh a nature to be of much value for forage. Like the rest of the species of this genus, the seeds, with their adherent awns, are very troublesome to sheep. Seeds ripen in October and November.

ARISTIDA LEPTOPODA, Benth.
"Three-awned spear grass."

A perennial species, found on rich soils in the interior, where it generally grows about 2½ feet high. Whilst young, it affords a capital forage for stock, but when it becomes old, is of a dry, wiry nature, and seldom eaten. It is a grass, however, that much improves by cultivation, and loses a great deal of the natural harshness common to uncultivated ones. The seeds and adherent awns are very troublesome to sheep. Seeds during November and December.

ARISTIDA RAMOSA, R. Br.
"Three-awned spear grass."

A perennial species, found both on rich and poor soils, generally over a greater portion of the Colony. Not of much value for forage, and, like the rest of the species of the genus, the seeds, with their adherent awns, are very troublesome to sheep. Seeds in October and November.

ARISTIDA STIPOIDES, R. Br.
"Three-awned spear grass."

A perennial species, found on poor soils and sand ridges in the interior, where it generally grows from 1 foot to 2 feet high. It is the worst of all the species for forage, as most of the plant is occupied by its inflorescence. The seed awns are very long, and most troublesome, both to wool and the animals' eyes. It ripens its seeds in November.

ARISTIDA VAGANS.

" Three-awned spear grass."

A slender, perennial, much-branched species found principally on poor stony rises in the coastal districts. It is quite common near Rookwood, and a short way out from North Sydney. During the winter and early spring months it makes a fair amount of leafy bottom, which affords some forage for stock ere many of our indigenous grasses start into growth. The awns of this species are much shorter, and consequently less troublesome than most of the other species of the genus. Seeds ripen during October and November.

ARTHRAXON CILIARE, Beauv.

" Swamp grass."

A perennial straggling species, with much-branched stems, sometimes rising a foot high. The leaves of this grass are nearly ovate in outline, and about 1¼ inches long. It is usually found growing in or near swamps, and, as far as it is at present known, only in the New England district in this Colony. It can hardly be classed as a valuable forage grass on account of its rarity, although cattle may occasionally be seen wading in the water to browse upon it. This grass does not produce much seed; whatever little there is, however, ripens in January.

ARUNDINELLA NEPALENSIS, Trin.

" Nepal grass."

A tall, glabrous, perennial species, in the warmer portions of the Colony attaining a height of 6 to 8 feet. In the cooler parts, however, it rarely exceeds half that height, and I have had specimens from near Windsor only a foot high. As might be supposed in such a variable grass as this one is, the forage it yields is not always of the same nature, when grown in different districts. In the colder parts of the Colony it is generally of a harsh nature, but in the warmer districts it yields capital forage, much relished by cattle. Under cultivation it is much improved, and if cut directly it shows its flowering stems it makes capital hay. The seeds ripen in November, December, and January.

ASTREBLA ELYMOIDES, Bail. et F.v.M.

" Mitchell grass."

A perennial species, of rather straggling habit, which, until quite recently, was only known to grow in the Warrego district of Queensland. A short time ago I had specimens sent for identification from our North-western districts. This grass has undoubtedly escaped the observation of many collectors, from the fact that the flower spikes often lie prostrate on the ground. It is a most excellent forage grass, and is held in great esteem amongst stock owners. Its thick, wiry roots penetrate the ground to a great depth, which enables the plant to withstand the most protracted drought. It is one of those grasses that sprout prolifically from every joint, when there is a slight rainfall after a long spell of dry weather. The seeds of this grass are like small grains of wheat, which at one time were used by the aborigines as an article of food. The seeds ripen in November and December.

ASTREBLA PECTINATA, F.v.M.
"Mitchell grass."

A perennial species, growing from 1½ to 3 feet high. On rich chocolate soils it grows into large tussocks, and produces a great amount of rich succulent herbage, which is much relished by all herbivora. Pastoralists in the western districts speak very highly of this grass, both for its drought-enduring qualities and its fattening properties. In dry seasons, when other feed is scarce, cattle may often be seen licking the broken parts of this grass from the ground, and they seem to fatten on it, even when it is in a very dry state. Although its natural habitat is purely western, it will grow equally as well in the coastal districts. This I have proved by cultivating it on the eastern side of the Dividing Range. The thick, wiry roots of this grass penetrate the ground to a great depth, which enables the plant to withstand the most protracted drought, and for this reason it is a most valuable standby for the pastoralist. An experienced drover once told me that stock would travel further and keep in better condition when fed on this than on any other grass in Australia. When cut just as the flower spikes appear it makes excellent hay, and if left growing a little longer would make good ensilage. The seeds of this grass, when ripe, are like small grains of wheat, and at one time they formed an article of food to the aborigines. The seeds ripen in November and December.

ASTREBLA TRITICOIDES, F.v.M.
"Mitchell grass."

Flora Austr., vol. VII, p. 602: An erect glaucous grass, very near *A. pectinata*, the leaves more or less scabrous or ciliate on the edges; spikes 3 to 6 inches long; spikelets alternate, not closely imbricate, and often almost erect and at some distance from each other. Outer empty glumes usually very unequal, the lowest short, the second four or five lines long. Flowering glumes shorter, the lateral lobes shorter and more rigid than in *A. pectinata*, and the awn much exceeding them, the dorsal hairs appressed and silky. Reference to plate :—A, spikelet ; B, floret ; C, grain, back and front views, variously enlarged. See frontispiece.

A perennial grass, usually a taller growing plant than the last species, and the flower spikes are often more than 6 inches long. On rich soils it produces a great amount of rich herbage which stock of all kinds are remarkably fond of. Cattle will fatten on this grass even when it is much dried up, during drought time. If it is cut when it first shows signs of flowering it will make excellent hay, and if left a little longer should make good ensilage. I have had this grass under cultivation and can thoroughly recommend it to be sown for permanent pasture either in the coastal or western districts. Before it is sown in the former place, however, the land must be thoroughly drained, if not naturally so situated, for this grass is very impatient of too much moisture. The seeds when ripe are like small grains of wheat, and at one time formed an important article of food to the aborigines. There is a variety of this grass called *lappacea* (Danthonia lappacea, of Lindley), which I have often recommended to be cultivated for the grain it yields. These grains are like small grains of wheat, and they separate most easily from the chaff. The ears (which are often more than 6 inches in length) are like large wheat ears, and where the latter would not grow, owing to great climatic heat, the former might, after a few years of careful cultivation and selection, be found an excellent substitute. The grain of this grass was at one time largely used by the aborigines as an article of food. This species and its variety ripen their seeds during October, November, and December.

BROMUS ARENARIUS, Labill.
"Brome, oat, or barley grass."

An annual species, growing from 1 foot to 2 feet high. A valuable grass to have in pastures, as it makes all its growth during the winter and early spring months, when many of our indigenous grasses are dormant. This grass is looked upon with great favour by stock-owners, especially on our far western plains, herbivora of all kinds being remarkably fond of it. It is much improved by cultivation, and, when grown on a good soil, yields a great amount of rich succulent herbage. The seeds ripen during September and October. Habitat: generally all over the Colony.

CENCHRUS AUSTRALIS, R. Br.
"Burr-grass."

A perennial species found growing on hill sides and on low scrub lands in various parts of the Colony. In the latter place it grows into large tussocks which are often 9 feet high. The herbage of this coarse-growing grass is very rough to the touch, and cattle seldom or never eat it, except after a fire, when the herbage is tender and succulent for a few weeks, and consequently they will eat of it. Sheep should never be depastured on land where this grass grows, and more especially when near ripening its seeds, for they are often troublesome to get out of the wool. The seeds ripen during November and December. This grass might be utilised for binding riverbanks or similar places against the fury of flood-waters, for its tough fibrous roots penetrate the soil to a great depth.

CHAMÆRAPHIS PARADOXA, Poir.
"Swamp couch."

A semi-aquatic species, growing in or near swamps, principally in the coastal districts. As the water recedes it forms a good sward of succulent herbage, which cattle are fond of, and they may occasionally be seen wading in the water to browse upon its floating stems. The seeds ripen during the summer months.

CHAMÆRAPHIS SPINESCENS, Poir.
"Floating couch grass."

A semi-aquatic grass, which, when growing in water, forms large floating masses. As the water recedes the creeping stems root into the soft mud and form a good sward of succulent herbage, which horses and cattle are remarkably fond of. In the western districts I have seen horses wading nearly up to their bellies in water to browse on this grass. Found near swamps in various parts of the Colony. The seeds ripen during the summer months.

CHLORIS ACICULARIS, Lindl.
"Umbrella or spider grass."

A glabrous erect perennial species, growing from 1 foot to 2 feet high. This grass grows plentifully in sandy and loamy soils in the interior. Its strong fibrous roots penetrate the soil to a great depth, which enables it to withstand the most protracted drought. During the summer months it yields a great amount of nutritious herbage, which is much relished by all herbivora. If cut when it shows its flower-stems it makes capital hay. This grass is well worthy of extensive cultivation in the arid parts of the interior where it may not already be growing. It produces a great amount of seed, which germinates readily under ordinary conditions, so no great difficulty is in the way of its dissemination. The seeds ripen in November and December.

CHLORIS TRUNCATA, R. Br.

"Star or windmill grass,"

A most variable species as regards height and size of inflorescence. In some situations it grows only 6 inches or a foot high, with the inflorescence only 4 inches across it. In other situations it will grow 3 feet high with the inflorescence a foot across it. This grass is generally found growing all over the Colony, and on rich alluvial soils it produces a great amount of succulent herbage much relished by all herbivora, sheep being particularly fond of it. I have had this grass under cultivation, and the bulk of herbage it yields is enormous. If cut when the flower-stems appear, it can be made into splendid hay. It is a perennial grass and its seeds ripen in October and November.

CHLORIS VENTRICOSA, R. Br.

"Blue-star grass."

An erect perennial species, which grows from 2 to 3 or more feet high on good soils. During the summer months it produces a fair amount of rich succulent herbage, which is much relished by all herbivora. If cut when it begins to flower it makes capital hay. This grass would well repay cultivation, and as it produces an abundance of seed it could easily be collected and disseminated in various parts of the country. The seeds ripen in October and November.

CHRYSOPOGON GRYLLUS, Trin.

"Golden beard."

An erect glabrous perennial grass, growing from 2 to 4 feet high. It is principally found in the interior; and on rich soils it grows into large tufts. During the summer months it makes a large quantity of succulent herbage, which is greatly eaten by all herbivora. Towards the end of summer the flowering stems become hard and cane-like, then stock seldom eat it. This prolific grass might very well be grown for ensilage, or if cut directly it shows its flower stems it would make good hay. The seeds ripen in November and December.

CHRYSOPOGON PARVIFLORUS, Benth.

"Scented golden beard."

This species and its varieties are easily known by the peculiar fragrance of its flowers on being rubbed in the hand. A tall growing perennial grass of 2 to 3 feet, and forming large tussocks on rich soils. It yields a great amount of herbage, and when it is in a young state is much relished by stock, but on becoming old it is very harsh and seldom eaten. Under cultivation this grass yields a great amount of succulent herbage, and if cut when the flowers first appear it makes capital hay, and if cut a little later on it should make good ensilage. This grass is generally found in the coastal districts. The seeds ripen in November and December.

CYNODON DACTYLON, Pers.

"Couch grass, Doub grass, Bermuda grass."

A perennial species with prostrate stems often creeping, and rooting at every joint. When it gets thoroughly established on good soils, however, the stems will grow from 1 foot to 2 feet high, if left unmolested for a time. In the coastal districts, where the frost is not too severe, it is the best native grass we have for making lawns. It is also valuable for consolidating earth banks, binding loose sand, and protecting river banks against the fury of flood waters. This grass should never be sown or planted except in places where it is required to remain permanently, for its numerous underground stems are most difficult to eradicate if they get into cultivated land. The drought enduring qualities of this grass are something remarkable, and if it once gets well established in the soil it is neither affected by very dry weather, nor close grazing, nor from being constantly trampled upon by stock. It is a most valuable pasture grass, which herbivora of all descriptions eat greedily of and fatten on. When grown under close paddocking, three crops may be cut in one season, and it makes splendid hay. Animals will thrive on the underground stems of this grass. Baron Von Mueller and L. Rummel give the following chemical analysis made on the very early spring growth of this grass: Albumen, 1·60; gluten, 6·45; starch, 4·00; gum, 3·10; sugar, 3·60 per cent.

DANTHONIA BIPARTITA, F.v.M.
" A Mulga grass."

A perennial species growing from 1 foot to 2 feet high. The stems rise from almost bulbous often woolly bases, which probably act as storage reservoirs to the plant, for it withstands the most protracted drought in what is often termed the desert interior, and its light green leaves may often be seen when the surrounding vegetation is somewhat dried up. It is a very nutritious and much esteemed grass, which herbivora of all kinds are fond of. It is found only in the arid interior, where the mulga (Acacia aneura, F.v.M.) tree grows, and in consequence stockmen have given it the common name of those trees. The seeds ripen in October and November.

DANTHONIA CARPHOIDES, F.v.M.
" Oat grass."

A perennial species, rarely exceeding 1 foot high. It is not found plentifully anywhere in the Colony, so can hardly be classed as a valuable forage grass, but what there is of it stock eat down with avidity. This grass is generally found in the colder parts of the Colony. The seeds ripen during November and December.

DANTHONIA LONGIFOLIA, R. Br.
" White-topped grass."

A perennial species, growing from 1 foot to 3 feet high. A superior pasture grass when found on the rich alluvial flats in the coastal districts, but of a hard wiry nature when found growing under less favourable circumstances. This grass is much improved when under cultivation, and yields a great amount of nutritious herbage, much relished by all herbivora. It will also make capital hay if cut directly the flower-stems appear. The roots of this grass penetrate deeply into the earth, which enables it to withstand a great amount of dry weather. Generally found in the coastal districts and on the high table-lands of the Colony. It is a prolific seed-bearing grass, and the seeds ripen during October, November, and December.

DANTHONIA PALLIDA, R. Br.
" Silver grass."

A perennial species, growing from 2 to 3 feet high. Generally found on rich soils, both in the coastal and western districts. During the summer months it yields a great amount of rich succulent herbage, which is greedily eaten by stock of all kinds, sheep being particularly fond of it.   Like many other species of this genus of grasses, it produces an abundance of seed, which germinates readily after showery weather in spring-time, and in consequence it has withstood the overstocking of runs much better than many other grasses, for in some situations it is still fairly plentiful.   This species would well repay cultivating for hay.   In the interior this grass ripens its seeds in October and November.   In the coastal districts it is generally one month later.

DANTHONIA PARADOXA, R. Br.

" Curious oat grass."

A perennial grass, growing from 2 to 3 feet high, and found in the coastal districts north of Port Jackson.   It does not appear to be plentiful anywhere, so very little is known of its qualities as a forage plant.   The seeds ripen in November and December.

DANTHONIA PAUCIFLORA, R. Br.

" Meagre-flowered oat grass."

A perennial alpine grass, rarely exceeding 6 inches in height.   A good grass for high altitudes, and, although it does not produce much forage in some situations, it forms a beautiful thick sward.   Seeds in January and February.

DANTHONIA PILOSA, R. Br.

" Hairy oat grass."

A perennial species, growing from 1 foot to 2, or even more feet high. An excellent pasture grass, which yields a quantity of rich succulent herbage near the base, and it is greedily eaten by stock of all descriptions.   It is not particular as to soil or situation, for it grows equally as well in the coastal districts as in the interior, and is highly spoken of by pastoralists as a good summer grass.   This species is much improved by cultivation, and it makes capital hay.   It is not likely to be lost by overstocking, as is the case with many other grasses, for it is a prolific seed-bearer, and the seeds germinate readily in showery weather in the autumn and spring months.   The seeds ripen in October and November.

DANTHONIA RACEMOSA, R. Br.

" Racemed oat grass."

A perennial slender grass, growing from 1 foot to 2 feet high, and found in the coastal and colder districts of New South Wales.   There are two well-marked varieties of this species, but with the exception of a slight difference in the inflorescence, their qualities are much the same.   All of them are most excellent pasture grasses, affording a very good forage for sheep. The qualities of these grasses do not appear to be affected by growing in different soils, for in all parts of the Colony where they grow stock eat them with avidity, and thrive on them.   They produce an abundance of seed, which ripens in the coastal districts in October and November, and in the colder parts one month later.

DANTHONIA ROBUSTA, F.v.M.

" Robust oat grass."

A perennial species found on the southern mountains in New South Wales. The stout stems rise from a thick horizontal root to 4 or 5 feet in height, although this is the largest species of the genus found in Australia, and forms large patches of foliage, which, at a casual glance, may look very coarse, still it is not despised by stock, and even small herbivora will eat it when in a young state, for then the herbage is rich and succulent. This valuable alpine grass is worth disseminating throughout the colder parts of the Colony, where such vegetation may be scarce. The seeds ripen in December, January, and February.

DANTHONIA SEMIANNULARIS, R. Br.
"Wallaby grass."

A perennial species of variable habit, sometimes only 6 inches high, at other times rising to 3 or more feet. In all its varied forms, however, it is one of the most nutritious grasses in the Colony, and unlike most other species of this genus, it will grow more or less all the year round. Stock of all descriptions are remarkably fond of it, and crop it so close down that in the colder parts of the Colony it gets little chance to perfect any seed. In the warmer parts, however, it produces an abundance of seed, which germinates readily after showery weather in the autumn or spring months. The roots of this grass penetrate deeply into the ground, which enables the plant, when growing in the interior, to withstand long spells of dry weather with impunity. Under cultivation, this grass produces a great amount of rich succulent herbage, which makes splendid hay. It would well repay systematic cultivation either for permanent pasture or making into hay. Although this grass is not particular as to soil or situation, still it grows best on a moderately rich strong loam, of good depth. In the interior this grass ripens its seeds in October, but in the coastal districts and colder parts of the Colony it is generally one or two months later.

DESCHAMPSIA CÆSPITOSA, Beauv.
"Tufted hair grass."

A perennial grass, growing from 1 foot to 3 feet high, and is generally found on low marshy land, or in wet places on mountain sides, in the southern portions of the Colony. Except in a young state, stock seldom touch it. As regards its nutritive qualities, it will be seen from the Woburn experiments that its cultivation cannot be recommended. "At the time of the seed ripening it yielded at the rate of 10,209 lb. of green produce per acre, which lost in drying 6,891 lb., and afforded of nutritive matter only 319 lb." Johnson, in his work on British grasses, says of the tendency of this grass to form tussocks, "In the economy of nature, these tufts, so unsightly and disfiguring to the cultivated landscape, are valuable by contributing to elevate and solidify low lands liable to be overflowed by rivers, and where they occur on hill and mountain slopes, by binding the spongy soil and preventing the slips which would leave them bare." The seeds ripen during November and December.

DEYEUXIA BILLARDIERI, Kunth.
"Bent grass."

A species growing from 6 to 18 inches high, according to the soil and situation it is found in, of perennial growth when found in moist pastures, but on high dry land it dies on the approach of hot weather. On good soils it yields a fair quantity of rich succulent herbage, of which sheep are very fond. This grass has an extensive range of growth in the coastal districts, being found from Illawarra to the Tweed. It produces a quantity of seed which ripens in October and November.

DEYEUXIA BREVIGLUMIS, Benth.

"Bent grass."

A slender perennial (?) species, growing from 1 foot to 1¼ feet high. So far as is at present known this grass has not a wide range of growth in this Colony, being found only on high table-lands of New England. During the summer months it yields a fair amount of rich herbage much relished by all herbivora. It does not produce much seed, what little there is, however, ripens in November.

DEYEUXIA FORSTERI, Kunth.

"Bent grass."

An annual species growing from 1 foot to 2 feet high. On rich soils it produces a great amount of rich succulent herbage, which is greedily eaten by all herbivora. This grass makes most of its growth during the winter and early spring months, and is a most valuable addition to our pastures when most of our native grasses are dormant. It is much improved by cultivation, and if cut when the flower panicle first appears it makes capital hay. This grass is generally found growing all over the Colony, and as it produces an abundance of seed, there would be no difficulty in collecting sufficient to sow large areas in any district required. During October, November, and December, the ripe, large panicles are blown in all directions, and I have seen, on the arrival at Sydney Railway Station of an up-country train, some of these large panicles stuck to the lower part of the carriages. The seeds ripen in October and November. Baron Von Mueller and L. Rummel give the following chemical analysis of the spring growth of this grass: Albumen, 4·08 ; gluten, 8·81 ; starch, 1·34 ; gum, 2·50 ; sugar, 9·75 per cent.

DEYEUXIA FRIGIDA, F.v.M.

"Bent grass."

A perennial grass with long, weak stems, found only on mountains in the southern parts of the Colony, in which situations it affords a fair amount of herbage during the summer months. Seeds in January and February.

DEYEUXIA MONTANA, Benth.

"Bent grass."

Another perennial species found on the southern mountains, but of upright growth, usually from 1 foot to 2 feet. In some places it is very plentiful, and affords tender herbage for sheep. Worth disseminating in mountainous districts. Seeds in January and February.

DEYEUXIA NIVALIS, Benth.

"Bent grass."

This in another perennial species found on the southern mountains, but it rarely exceeds 1 foot in height. In some places it is plentiful, and forms a good sward. During the summer months it yields a fair amount of herbage. Worth disseminating in cold, mountainous districts. Seeds during January and February.

DEYEUXIA PLEBEIA, Benth.

"Bent grass."

A slender, tufted annual grass, rarely exceeding 1 foot high. It is fairly plentiful in some situations in the coastal districts, but it cannot be regarded as good forage grass, for in some seasons it is of very short duration, withering off on the advent of hot weather. It does not produce much seed ; what little there is, however, ripens in October and November.

DEYEUXIA QUADRISETA, Benth.

" Bent grass."

An erect perennial species, very variable in stature, but usually from 1 foot to 3 feet high. It is not particular as to soil or situation, for it is found growing both on ironstone ridges and on rich alluvial flats, principally in the coastal districts, but also in New England, although not so plentifully. On rich soils this grass yields a great amount of forage, and while young it is fairly good feed for cattle ; when it becomes old, however, the stems become hard and cane-like, then it is seldom or never touched. This harsh grass is never at any time of any value as forage for sheep, neither would it make good hay. The seeds ripen in November, December, and January.

DEYEUXIA SCABRA, Benth.

" Bent grass."

A perennial species, with weak decumbent stems 1 foot to 2 feet long. It is found in the coastal districts north of Port Jackson, and also in New England, but it does not seem to be very abundant anywhere; I have found specimens occasionally near Manly. Wherever it is found stock eat it freely, still it can scarcely be considered a good forage grass. This grass produces a fair amount of seed, which ripens during October and November.

DICHELACHNE CRINITA, Hook.

" Long-hair plume grass."

A perennial species growing from 2 to 3 feet high, and when in flower is a prominent feature in the pastures. This grass is found on various soils in different parts of the Colony. In the hot, dry districts, it is harsh and somewhat scanty of foliage, but in the coastal districts its character is altogether changed. On good soils it produces a great amount of rich succulent herbage, which is greedily eaten by all herbivora. This grass will continue to grow throughout the year on good pasture land that is fairly well sheltered, and where frosts are not too severe. It is a good grass for the dairy farmer, and if cut when the flowers first appear it makes excellent hay. This grass is a prolific seed-bearer, so that there will be no difficulty in collecting any quantity by those desirous of doing so. The seeds ripen during October, November, and December.

DICHELACHNE SCIUREA, Hook.

" Short-hair plume grass."

A most variable grass as regards the arrangement of its inflorescence. Sometimes it is dense and spike-like, at other times it is very loose and somewhat spreading, and to the casual observer the extreme forms would look as if they belonged to distinct species. It is a slender perennial grass, usually growing about 1½ feet high, and is found on the eastern side of the Dividing Range, and also in the New England District. It is a very quick-growing, succulent grass, and a most valuable one to have in the pastures, as it makes considerable growth during the winter and early spring months, ere many of the indigenous grasses show much signs of life. If cut when it shows its flowers it makes excellent hay. This grass produces a fair amount of seed, which ripens in October and November.

DIPLACHNE FUSCA, Beauv.
" Brown-flowered swamp grass."

A glabrous annual species, growing from 2 to 3 feet high.   In the coastal districts it is generally found in brackish swamps ; in the interior in shallow pools of water, or in damp situations.   During the summer months it affords a rich succulent herbage, greedily devoured by cattle.   This grass is worth disseminating in swampy land, where hardly anything else would grow, as it would afford a valuable lot of herbage during a long spell of dry weather, when the surrounding country was dried up.   This species produces a quantity of seed, which ripens during January, February, and March, on swampy land, but in drier places it generally ripens in November.

DIPLACHNE LOLIIFORMIS, F.v.M.
" Rye-like grass."

A slender-erect annual (?) species, and when found growing in the western districts it rarely exceeds 8 inches in height, but some specimens I received from Albury were over a foot high.   It is generally found growing on light sandy soils, but it is not of much value for forage.   Sheep will occasionally eat it, but cattle often pull it out of the ground when grazing where it grows, as its roots take such a slight hold of the ground.   Produces plenty of seed, which ripens in October.

DISTICHLIS MARITIMA, Rfn.
" A sea-side couch grass."

A dwarf creeping species, valuable only for binding loose sand and embankments near the sea.   Found on the southern parts of the sea coast. It can be propagated by division of its roots.

ECHINOPOGON OVATUS, Beauv.
" Rough-bearded grass."

A rough-leaved grass, growing from 1 foot to 2 feet high.   It is generally found on poor soils and on hill-sides in the coastal districts, from Illawarra to the Tweed, also in New England.   This grass affords a fair amount of herbage during the winter and early spring months, which, however, is wanting in nutrition.   Cattle will occasionally browse upon it when other feed is scarce.   It does not improve very much under cultivation.   It is an annual, and seeds in September and October.

ECTROSIA LEPORINA, R. Br.
"Hare's-tail grass."

A glabrous, slender, perennial species, growing from 2 to 2½ feet high. It is only found in the north-western portions of the Colony, but it is not plentiful anywhere.   The roots of this grass penetrate deeply into the soil, which enables it to withstand the most protracted drought.   In the early stages of its growth it yields a rich succulent herbage, which is much relished by all herbivora.   When it becomes old, however, the stems are rather dry and wiry, and sheep will seldom touch it if other herbage is plentiful.   This grass is well worth disseminating on our arid plains, for it will thrive on the driest of soils.   If left unmolested for a time it produces a great quantity of seed, which ripens in October and November.   This species has a panicle from 3 to 6 inches long, with purplish spikelets, and long slender awns, giving it a very ornamental appearance when in seed. There is a small variety of this beautiful grass, rarely exceeding 6 inches in height, which is worth the attention of horticulturists. Both these grasses would make very ornamental objects in a garden.

ELEUSINE ÆGYPTIACA, Pers.
" Egyptian finger grass."

An annual species, of tufted growth, but sometimes with creeping stems rooting at every joint, the lateral growths rarely ever exceeding 1 foot in height. This grass is generally found all over the western interior, and in some situations it is fairly plentiful. During the hot summer months it yields a rich succulent herbage, which is much relished by all herbivora, sheep being particularly fond of it. It is very highly spoken of by all sheep owners as a most nutritious grass. I have had this species under cultivation, and can highly recommend it both for pasturage and hay. It produces an abundance of seed, so that there would be no difficulty in disseminating it throughout the Colony. This grass likes a good soil to grow in; seeds ripen in October and November.

ELEUSINE INDICA, Gærtn.
"Crow-foot or crab grass."

A coarse, erect, tufted perennial grass, growing from 1 foot to 2½ feet high, according to soil and situation. It is principally found in the coastal districts from the Hunter to the Tweed, and in some situations it is fairly plentiful. This grass may be recognised by its dark green colour, strong stalks, and digitate panicles, the spikelets of which are flat, and overlap each other. It grows all the year round, but during the summer months it yields a rich succulent herbage, much relished by cattle. It will make good hay. The tough fibrous roots of this grass penetrate deeply into the soil, and it is useful for binding the banks of rivers, dams, and loose earth. The seeds ripen in November and December.

ELYTROPHORUS ARTICULATUS, Beauv.
Jointed blue grass.

An annual species, rarely exceeding a foot and a half in height, and is found on the Murray, and at several stations in the interior, but so far as I have been able to find out, it is not very plentiful. During the early spring months it affords a fair amount of herbage which sheep are fond of; seeds in October.

ERAGROSTIS BROWNII, Nees.
" Brown love grass."

A perennial species growing from 1 foot to 4 feet high. There are several varieties of this most excellent pasture grass found in various parts of the Colony, both on rich and poor soils. It will grow all the year round, and on alluvial soils will yield a very great amount of succulent herbage, much liked by all pasture animals. It will also make splendid hay. The strong fibrous roots of this grass penetrate deeply into the soil, which enables it to withstand the most severe drought. Its herbage keeps beautifully green all through an ordinary dry summer. I have had this grass under cultivation, and the amount of herbage it yielded was something astonishing. I can very highly recommend it for cultivation, both to pastoralists and farmers. There should be no difficulty in bringing this grass under cultivation, for it

c

produces an abundance of seed when left undisturbed for a time. If it were growing any way plentifully in a reserved area, a boy could, when the seeds were ripe, collect two or three bushels a day, which would be enough to sow as many acres. The seeds ripen in November, December, January, February, and March.

ERAGROSTIS CHÆTOPHYLLA, Steud.

"A love grass."

A perennial species growing from 6 inches to 1 foot high, principally found on stiff loamy soils, in the arid interior. The stems rise from almost bulbous woolly bases, which no doubt act as storage reservoirs, and enable the plant to withstand the most prolonged drought. The herbage, however, is of too wiry a nature to be of much value for forage. It is only eaten during very dry times, and when other grasses fail. When this grass is in flower it is a most ornamental plant. It produces a quantity of seed, which ripens in November and December.

ERAGROSTIS DIANDRA, Steud.

"A love grass."

An erect-growing perennial species, the stems rising from 1 foot to 2 feet. This grass is found in the coastal districts, and also in New England. A valuable drought-resisting species, and during the summer months it yields a fair quantity of rich succulent herbage, which is greedily eaten by all herbivora. It also makes capital hay. This grass is worth cultivating, either for permanent pasture or hay. It produces plenty of seed, which ripens in November and December.

ERAGROSTIS ERIOPODA, Benth.

"A love grass."

A very pretty perennial species, growing about 1½ feet high. It is generally found on the clayey soils in the arid interior. This grass has remarkable drought-resisting powers, but its herbage is of a hard wiry nature, and is seldom eaten by stock unless other grasses fail. The seeds ripen in November and December.

ERAGROSTIS FALCATA, Gaud.

"A love grass."

A slender perennial species rarely ever exceeding 1 foot, and sometimes only a few inches high. It is mostly found on sandy soils in the arid interior. This grass may easily be recognised amongst other of the genus, by the one-sided arrangement of its panicles. I have had sent to me from the Darling and Lachlan Rivers some depauperate specimens of this grass, which, however, are only of interest to the microscopist. None of the forms of this grass are of any value to the pastoralist. It seeds in November.

ERAGROSTIS KENNEDYII, Tur. et F.v.M.

"Love grass."

This grass is named after Mrs. Kennedy, of Wonominta, near Wilcannia, who found it in the district last year. A slender, tufted, perennial grass of about 1 foot. Glabrous, except a few cilia at the orifice of the sheaths. As far as is at present known, it is only found on the red soils near Wilcannia, and a few other western stations in New South Wales, and over the Queensland border. Like some of its congeners, it is capable of withstanding great drought and heat. Very little is known as regards its value as a forage grass. It produces a fair amount of seed, but it is exceedingly small. The seeds ripen during the latter part of summer.

ERAGROSTIS LACUNARIA, F.v.M.
" A love grass."

A very slender perennial species, growing from 6 to 15 inches high, and generally found on sandy soils in the arid interior. The panicles are mostly of a dark purple colour, which gives quite a feature to the plant. The scanty herbage is of a hard wiry nature, and is not much eaten by stock, except when other herbage fails through drought. The seeds ripen in October and November.

ERAGROSTIS LANIFLORA, Benth.
" A love grass."

A perennial species, rarely ever exceeding 1 foot high. The wiry stems rise from thick, almost bulbous, woolly bases, but the herbage is not of any value for forage. Generally found on clayey soils in the remote interior. Seeds in October and November.

ERAGROSTIS LEPTOSTACHYA, Steud.
" A love grass."

A slender perennial species growing about 1½ feet high. It is found in the coastal districts from Illawarra to the Tweed; and also in New England. In some situations it is fairly plentiful, and on good soils it yields a rich succulent herbage much sought after by all pasture animals. This grass will grow on land where it is partially shaded with trees, and in such circumstances, will afford a tender herbage during the winter and early spring months. This grass is much improved by cultivation, and if cut when the flower stalks first appear, it makes good hay. It produces an abundance of seed, which ripens in October and November.

ERAGROSTIS MEGALOSPERMA, F.v.M.
" A love grass."

A rather tall perennial species growing from 2 to 3½ feet high. This grass is rare in the Colony, being found only, as far as at present known, in the northern coastal districts. Where this grass is found, it is generally an indication of rich land. In a young state, its herbage is succulent and is much relished by all pasture animals. It will stand a great amount of dry weather and remain green during the greater part of summer. This grass produces a fair quantity of seed, which ripens in November and December.

ERAGROSTIS NIGRA, Nees.
" A love grass."

A rather slender species, with dark-coloured spikelets arranged in a spreading panicle. This grass seems to be rare and restricted in its habitat, being only found, as far as is at present known, in the New England District. It does not produce much seed; what little there is, however, ripens in December.

ERAGROSTIS PILOSA, Beauv.

" A love grass."

An annual species growing from 1 foot to 3 feet high, which is generally found all over the Colony. This grass is not particular as to soil or situation, as it may frequently be seen growing both on stony ridges, and on rich meadow land. On good soils, however, it is a most prolific grass, and during the summer months affords a large amount of good herbage much relished by stock of all kinds. This grass will spring into growth at any time of the year after a shower of rain, and will afford winter feed where frosts are unknown. Under cultivation this grass yields a great amount of herbage which can be may into capital hay. It perfects a great amount of seed, which can be collected at any time during the summer and autumn months.

ERAGROSTIS TENELLA, Beauv.

" A love grass."

A delicate erect tufted annual grass, sometimes only 6 inches, at other times nearly 2 feet high. It is principally found on the north-western plains of the Colony, and, during the summer months, forms an excellent pasture grass for sheep. It will also make good hay. This grass is worth disseminating on all our arid plains especially where sheep are grazed. It is very easily recognised by its delicate reddish inflorescence, and as it produces a quantity of seed no difficulty would be experienced in collecting it. The seeds ripen in October and November.

ERIACHNE ARISTIDEA, F.v.M.

A many-branched perennial species, rarely exceeding a foot and a half in height. It is principally found on the arid plains of the interior where, however, it does not grow very abundantly. Whilst it is in a young state the herbage is much relished by sheep, but when it becomes old it is often hard and wiry, then it is seldom or never eaten, except in times of great scarcity. The seeds ripen in October.

ERIACHNE OBTUSA, R. Br.

A many-branched perennial grass of variable habit. Sometimes it is only a few inches high, at other times it exceeds 2 feet in height. This species is pecular to the back country where it grows on sandy soils, but not very abundantly. During the early summer months while the herbage is young, it is much relished by stock of all kinds, particulary by sheep. When the herbage becomes old, however, it is of a hard wiry nature, and it is seldom eaten except when other herbage is scarce. The drought-enduring qualities of this grass are something remarkable, for it does not seem to be much affected either by the extreme heat of summer, or the other extreme of cold in winter. This grass is worthy of dissemination on the arid sandy plains of the interior where very few other kinds of grasses could exist. It produces a fair amount of seed which ripens in October and November.

ERIOCHLOA ANNULATA, Kunth.

" Early spring grass."

A perennial species growing from 1 foot to 2 feet high. A superior pasture grass found in the coastal districts, and the colder parts of the Colony. It will grow and furnish feed nearly all the year round in the coastal districts, but during early summer months it yields a great amount of rich succulent herbage greedily devoured by stock of all kinds. If cut when the flowers appear it makes excellent hay. This grass is worth the attention of dairy farmers, as milch cows are fond of grazing upon its rich succulent herbage. When left undisturbed for a time it produces a great amount of seed which ripens in November and December.

ERIOCHLOA PUNCTATA, Hamilt.
" Early spring grass."

An erect perennial grass of 2 to 3, or even more, feet high. It is found growing over a greater portion of the Colony from the coast to the arid interior, and in some situations it is fairly plentiful. This grass is found growing on various kinds of soil, but the one that suits it best is a deep chocolate loam, where it will produce a rich succulent herbage, much relished by all herbivora. In sheltered situations in the coastal districts this grass will grow all the year round, but in the arid interior it only grows during the summer months. Its tough, fibrous roots penetrate the soil to a great depth, which enables it to withstand a very long spell of dry weather. During its growing period this grass is easily recognised by its glaucous appearance. Under cultivation it produces a great amount of herbage which can be made into excellent hay. When left undisturbed for a time it produces a great amount of seed, which ripens in November and December. In the interior the inflorescence of this grass is sometimes affected with a parasitic fungus—probably an ustilago.

FESTUCA BROMOIDES, Linn.
" Barren fescue."

An annual species of variable habit; sometimes only a few inches high, at other times over a foot in height. It is found generally all over the Colony, from the coast to the interior. It is very common on poor, sandy soils in the county of Cumberland during the winter and early spring months, but dies on the approach of summer. This grass is of no value for forage, nor for anything else in the economy of nature that I know of. It produces a great amount of seed, which ripens in September and October.

FESTUCA DURIUSCULA, Linn.
" Sheep's fescue."

An erect perennial species, growing from 1 foot to 2 feet high. This grass is found in the colder and mountainous parts of the Colony. It is adapted to a great variety of soils, but it prefers those that are light and rich, and under these conditions yield a fair amount of succulent herbage, of which sheep are particularly fond. This grass possesses the quality of withstanding, in a marked degree, the drought in summer and the cold in winter. Some forms of this species are often used in making lawns, and if well kept down with scythe or machine, in a short time will make a good sward. If left unmolested for a time this grass produces a fair amount of seed, which ripens in November and December, but in very cold places a month or two later.

Analysis of Festuca duriuscula.

Mr. Martin J. Sutton gives the following analysis of this grass.

	Grass in a natural state.	Dried at 212° Fahr.
Water	61·98	....
*Soluble albuminoids	·17	·44
†Insoluble albuminoids	1·50	3·94
Digestible fibre	6·53	17·18
Woody fibre	23·19	60·99
‡Soluble mineral matter	1·52	4·01
‖Insoluble mineral matter.....................................	·86	2·26
Chlorophyll, soluble carbo-hydrates, &c.	4·25	11·18
	100·00	100·00
*Containing nitrogen	·03	·07
†Containing nitrogen..	·24	·63
Albuminoid nitrogen	·27	·70
Non albuminoid nitrogen	·11	·29
Total nitrogen	·38	·99
‡Containing silica	·33	·99
‖Containing silica	·47	1·34

GLYCERIA FLUITANS, R. Br.
" Manna grass." -

A perennial species, with stems creeping in the mud, and sometimes floating on water; principally found in the coastal districts, but occasionally in New England and other cold districts. The floating or creeping succulent stems of this grass are much sought after by horses and cattle, and they may occasionally be seen wading in the water after them. The seeds are sweet and palatable, and ducks and other aquatic birds greedily feed upon them. The seeds ripen during the summer months.

GLYCERIA FORDEANA, F.v.M.
" Sweet grass."

An erect glabrous perennial species, growing from 2 to 3 feet high. It is generally found in moist situations about the Darling and Lachlan Rivers in this Colony. In dry seasons, when herbage is scarce, the succulent stems of this grass are much sought after by stock, and they often eat it so closely down that it is becoming scarce in some situations, where it once was plentiful. When left unmolested for a time it produces a quantity of seed which ripens during October and November.

GLYCERIA LATISPICEA, F.v.M.
" Sweet grass."

An erect species, the stems usually attaining a height of 2 or 3 feet. This grass has not a wide geographical range in the Colony. As far as I have been able to find out it does not occur out of the New England district. It is generally found in moist situations, but not in sufficient quantities to allow its qualities to be reported upon. Under ordinary circumstances it produces a fair amount of seed, which ripens in November and December.

GLYCERIA RAMIGERA, F.v.M.
" Cane grass."   " Bamboo grass."

A glabrous, rigid, perennial species, growing from 6 to 10 feet high, and usually forming large tussocks, in what are called " cane swamps " or "clay-pans," near the Murrumbidgee, Lachlan, and Darling Rivers, in this Colony. Although the stems of the grass are so hard and cane-like they send out a number of leafy branches near the top, which cattle greedily feed upon, but whether there is any nutriment in them I am unable to say. In the districts named this grass is often cut for thatching purposes, for which it is admirably adapted. It is said that roofs made of this grass will remain perfectly waterproof over twenty years. It seeds in November, December, and January.

HEMARTHRIA COMPRESSA, R. Br.
"Sugar grass."

A perennial species, with decumbent or creeping stems, often extending the length of 5 or 6 feet, the branches ascending to 1 foot or more. It is a purely coastal grass extending from Illawarra to the Tweed. I have found it in several places near Port Jackson, and often close to salt water. At the first glance this rather harsh wiry-looking grass would not impress any one as being a valuable forage, nevertheless it is a fact that stock of all kinds eat greedily of it. It is said that horses will leave all other herbage to browse upon this grass. It is generally found on low wet soils, or near swamps, and for covering such land hardly any other grass is more suitable, for in a very short time it forms a beautiful green sward. Its seeds ripen during the summer months. This grass can be easily propagated by division of its stems.

HETEROPOGON CONTORTUS, Rœm et Schult.
"Bunch spear grass."

A perennial species, growing from 1 foot to 3 feet high. It is found in the north-eastern parts of the Colony, and in some places it is fairly plentiful. A strong-growing leafy grass, affording a great amount of herbage, suitable for a cattle run, but when in seed is dreaded by the sheep owner on account of its numerous barbed seeds, which not only injure the wool, but often penetrate the skin and intestines of these animals. I have had this grass under cultivation and it yielded a great amount of rather coarse herbage, which, however, might make good ensilage. The tough fibrous roots of this grass penetrate deeply into the soil, and I can recommend it for binding river banks against the fury of floods, the banks of dams, and any loose earth. It seeds in November and December.

HIEROCHLOA REDOLENS, R. Br.
"Sweet-scented sacred grass."

A perennial species, growing from 2 to 4 feet high, which is generally found in moist places on the southern mountains. This grass is worthy of dissemination on cold, wet, marshy land, or on poor sandy soils near the sea, which it would help to bind with its long tough fibrous roots. There is some diversity of opinion as regards the nutritive qualities of this grass. However this may be, the succulent herbage is much sought after by all herbivora, and we must consider them the best judges at discriminating between good and bad forage. Moreover, this grass is particularly valuable for its fragrance (the active principle being coumarine), and if mixed with damaged hay cattle will eat it, although they may previously have rejected it. If left unmolested for a time it produces a great quantity of seed, which ripens during February and March.

HIEROCHLOA RARIFLORA, Hook.
" Sweet-scented grass."

A perennial species, growing from 2 to 3 feet high, which is principally found about Twofold Bay, in the southern parts of the Colony. It has nearly all the characteristics of the last-named species (to which it is closely allied), and can be used for precisely the same purposes, and, further, some of our southern colonists stuff their beds and pillows with the dried leaves of this sweet-scented grass, being under the impression that it induces sleep, and keeps away house vermin. This species does not produce a great amount of seed, what little there is, however, ripens in January and February. It can also be propagated by division of its roots.

IMPERATA ARUNDINACEA, Cyr.
" Blady grass."

A stiff-erect perennial species, growing from 1 foot to 3 feet high, and generally found in wet localities all over the eastern portions of the Colony. It is a valuable grass for binding the littoral sands, as its underground stems form a perfect net-work, which is most difficult to eradicate. It can also be recommended for binding river banks, the sides of dams, and any loose earth, but nowhere near cultivation. If it were ever allowed to get established on good land it would become almost irrepressible, for every small joint of its underground stems that is left in the ground will develop into a plant. In some instances this grass covers large areas of wet and often sour land, and if burnt off in October or November it will yield a capital herbage during the greater part of summer, which cattle eat with avidity. When this grass becomes old, however, it is very tough and harsh, and cattle seldom or never touch it whilst any other herbage is to be obtained. This grass has often proved a valuable stand-by for stock during prolonged droughts, especially after being burnt off in spring time, and I have known of instances where flocks and herds almost depended upon this species alone for forage during very dry times. This grass is easily recognised by its silvery white spike-like panicles, which are often 6 inches long. It is easily propagated by division of its roots.

ISACHNE AUSTRALIS, R. Br.
" Swamp millet."

A perennial species, with rather slender creeping stems, which root at the lower joints and ascend to about 1 foot. It is mostly found on wet land in the coastal districts, and also in New England. During the summer months it makes rapid growth, and its tender herbage is much sought after by all herbivora. This grass is a general favourite with pastoralists. It is worthy of dissemination on wet land all over the Colony, and, as it produces an abundance of seed under ordinary circumstances, there would be no difficulty in collecting quantities of it. The seeds ripen in November and December.

ISCHÆMUM AUSTRALE, R. Br.
" Southern grass."

A perennial species, growing from 2 to 3 feet high, and is principally found on moist soils in the coastal districts, from Port Jackson to the Tweed. This grass has strong underground rooting stems which enables it to withstand a lengthened period of drought, and it will continue to grow long after many others have died off. It is a valuable grass for binding the banks of rivers or dams, and any loose earth. To render this grass of any value for forage purposes it should be burnt off annually in October or November, after which it will produce a quantity of good herbage during the greater part of summer. When this grass becomes old the herbage is hard and wiry; then it is seldom or never eaten by stock. It produces a fair quantity of seed, which ripens in December and January. This grass can also be propagated by division of its roots.

ISCHÆMUM CILIARE, Retz.
"Hairy southern grass."

A very slender species, rarely exceeding 1 foot high, and only found in the coastal districts, but nowhere plentiful. Sometimes it grows in small tufts, at other times it forms a good sward. It is a capital pasture grass for sheep, and it retains its green appearance during the greater part of summer. It makes a good lawn grass, and it is easily kept in order. It does not produce much seed. What little there is, however, ripens in October and November. This grass can be propagated by division of its roots at any time during the spring of the year.

ISCHÆMUM PECTINATUM, Trin.
"Comb-like southern grass."

A perennial species, found growing on various soils in the coastal districts. On poor land it grows in dense tufts, but on fairly good soils it forms a splendid sward, which keeps its vivid green appearance during a greater part of the year. Having grown it as a lawn grass I can highly recommend it to be extensively planted, for it is easy to keep in order. This grass can be easily propagated by division of its roots at any time during the early spring months. It produces a fair amount of seed, which ripens in November and December. It is worthy of dissemination in the coastal districts as a pasture grass, for sheep are particularly fond of it.

ISCHÆMUM TRITICEUM, R. Br.
"Southern wheat grass."

A coarse, erect perennial species, growing from 2 to 3 or more feet high, found only in the north-eastern parts of the Colony. It is a valuable grass for binding the banks of rivers and dams, and also any loose earth. The strong underground stems form a perfect net-work, which are not easily eradicated. To render this grass of any value for forage purposes it should be burnt off annually in October or November, after which it will produce a quantity of good herbage during the greater part of summer. If cut when the flower-stems appear, it ought to make good ensilage. After this grass has flowered, its herbage gets very harsh and wiry, and stock seldom or never touch it. It produces a fair amount of seed, which ripens during January, February, and March. This grass can also be propagated by division of its roots any time during the early spring months.

LAMARCKIA AUREA, Mœnch.
"Golden-headed grass."

A most elegant tufted annual species, rarely growing more than 6 inches high, and found only in the arid interior, but nowhere very plentifully. It is of no value for forage purposes, on account of its fugitive nature, and, besides, the beautiful golden inflorescence occupies more than half the plant. It is worthy the attention of horticulturists, however, who might introduce it into gardens with advantage. This grass grows best on a light dry sandy soil. It produces a fair quantity of seed, which ripens in October.

LAPPAGO RACEMOSA, Willd.

"Small burr grass."

An annual species, spreading on the ground or ascending to from 6 inches to 1 foot in height. It is generally found growing all over the Colony, from the coastal districts to the arid interior, and in some situations it is fairly plentiful. This grass is not particular as to soil or situation, for it may be seen as often growing on dry stony ridges as in the more fertile pastures. During the winter and early spring months it produces a fair amount of herbage which sheep are fond of. After heavy rains in the summer months I have seen this grass spring up and grow quite luxuriantly in the month of January. Under ordinary circumstances the seeds of this grass ripen in October. Although it is one of the burr-seeded grasses, it is said to do no harm to the wool.

LEERSIA HEXANDRA, Swartz.

"Native rice grass."

A glabrous perennial species, always found in or near water in the north-eastern parts of the Colony. The stems root in the mud at the lower joints, and then often branch out several feet in height, or lie on the surface of the water. Stock are particularly fond of its succulent herbage, and they may often be seen wading in the water to browse upon it. This grass is worthy of dissemination near all water-courses in the eastern parts of the Colony; besides producing valuable forage during the summer months, it yields a fair-sized grain, somewhat like rice, which might be much improved by judicious cultivation and careful selection of varieties. It is also worthy of attention where water-fowl are kept, for they are particularly fond of the grain of this grass. When left unmolested for a time it produces plenty of seed, which ripens in December and January.

LEPTOCHLOA CHINENSIS, Nees.

"Weeping grass."

A glabrous perennial species, creeping and rooting at the base, the stems usually attaining a height of 2 or 3 feet. It is generally found along water-courses in the north-eastern parts of the Colony, and also in New England. During the summer months it produces a large amount of forage, which is much relished by all herbivora. If cut when the flower-stems first appear it should make good hay. This grass will be found valuable in binding the banks of rivers and dams, or any loose earthwork in damp situations. Its tall but very delicate and drooping grey-coloured panicles are very ornamental, and might with advantage, for scenic effect, be planted on the banks of artificial lakes, or in low damp places in pleasure grounds. This grass produces an abundance of seed, which ripens in November and December.

LEPTOCHLOA SUBDIGITATA, Trin.

"Cane grass," "Finger grass."

An erect, rigid perennial species, growing from 4 to 5 or more feet in height. It has usually a glaucous appearance, and is generally found growing on river-banks and around lakes all over the interior, and in some places it is fairly plentiful. It produces a great quantity of coarse herbage, and during the early summer months is much relished by stock. When it becomes old, however, it is hard and cane-like; then stock seldom or never touch it. If burnt off annually in October or November it yields a quantity of herbage during the rest of the summer. This grass produces a fair amount of seed, which ripens in November, December, and January.

LEPTURUS CYLINDRICUS, Trin.
" Salt marsh grass."

A tufted or slightly branching annual grass, rarely ever exceeding 1 foot in height. It is generally found on salt marshes in the southern parts of the Colony, but nowhere very plentiful. It is of little value, however, either for forage or for forming a sward in salt marshes. It does not perfect much seed; what little there is, however, ripens in January and February.

LEPTURUS INCURVATUS, Trin.
" Salt marsh grass."

A tufted or branching annual species, usually about 6 inches, but sometimes nearly a foot high. It is found growing on salt marshes near the Parramatta River and in the southern coastal districts, but nowhere very plentifully. It is of very little value for any purpose. This grass does not produce much seed; what little there is, however, ripens in January and February.

MICROLÆNA STIPOIDES, R. Br.
" Meadow rice grass."

An erect-growing perennial species, generally found in the coastal districts, but also on the Blue Mountains, and in New England. It usually grows from 1 foot to 2, but occasionally may be seen 3 feet high. The stems rise from a rather thick rhizome, and the roots penetrate deeply into the soil, which enables the plant to withstand a very long spell of dry weather. It is a most superior pasture grass, and in some situations will grow all the year round. It may easily be recognised during the winter and early spring months by the vivid green appearance of its foliage, which, in an ordinary season, it will retain throughout the summer. This grass produces a rich, succulent herbage, which is greedily sought after by all herbivora. If cut when the flower stems first appear, it makes excellent hay. This grass is much improved by cultivating it on good soils, and even during an ordinary season will produce a bulk of herbage that is quite astonishing. I can highly recommend this grass for permanent pasture (even under close feeding it will maintain a close turf), or for hay. There would be no difficulty in bringing this grass under cultivation, for in ordinary circumstances it produces a great amount of seed, which ripens in November and December.

Baron von Mueller and L. Rummell give the following chemical analysis made on the spring growth of this grass: Albumen, 1·66; gluten, 9·13; starch, 1·64; gum, 3·25; sugar, 5·05 per cent.

NEURACHNE ALOPECUROIDES, R. Br.
" Mulga grass."

A perennial species, usually growing from 1 foot to 1¼ feet high, and generally found in the south-western parts of the Colony, but not very plentifully. During the summer months it produces a fair amount of herbage which ·

sheep are fond of. It is worthy of wide dissemination on the south-western plains, for it affords herbage when many other grasses fail during drought-time. This may be accounted for by the fact that its strong fibrous roots penetrate deep into the soil, which enables it to withstand a long spell of dry weather with impunity. Its seeds are borne on ovoid or oblong spikes rarely exceeding 1 inch long. As a general rule it does not perfect a great amount of seed. What little there is, however, ripens in October, November, and December.

NEURACHNE MITCHELLIANA, Nees.

" Mulga grass."

So called from its frequently being found growing under or near the Mulga scrub (Acacia aneura) in the interior. A perennial species rarely exceeding 1 foot high, the stems of which rise from a thick woolly rhizome, which probably acts as a storage reservoir to the plant in very dry weather, for it will withstand and remain green throughout a very severe drought, even when growing on poor soils. This grass is a general favourite amongst pastoralists, as it affords a fair amount of forage, during adverse times, of which sheep are very fond. It is worthy of being disseminated all over the arid interior, and also on any dry, sandy, or sterile land. This grass produces a fair amount of seed, which ripens in September and October.

NEURACHNE MUNROI, F.v.M.

" Mulga grass."

This species grows in somewhat similar situations to the one described last. A perennial grass growing from 6 inches to 1 foot high, the stems of which rise from a more or less woolly or knotty base. It is a very rare grass, and found only in the back country. It does not produce much herbage, but what little there is, sheep eat freely. This grass produces a fair amount of seed, which ripens in October and November.

OPLISMENUS COMPOSITUS, Beauv.

" Slender panick grass."

A slender perennial hairy grass, with decumbent or creeping stems, always found growing under dense shade, and principally in scrubs in the coastal districts. It is not of any value for forage, as stock seldom or never touch it. It is a useful grass, however, for covering ground under the shade of trees, and in a short time forms a fine sward. There is a beautiful variegated form of this grass which may sometimes be found cultivated in gardens. It is easily propagated by pieces of its stems, as also from seed, which it bears in fair quantities. The seeds ripen during the autumn months.

OPLISMENUS SETARIUS, Rœm. et Schultz.

" Slender panick grass."

A perennial species, which grows in somewhat similar places to the one described last. It is a much more slender-growing grass than the previous one, and, like it, is of little or no value for forage, although it is a useful grass for covering ground under the dense shade of trees, and might be used for such a purpose with pleasing effect. It can be propagated. by pieces of its stem, as also from seed, which it bears in fair quantities. The seeds ripen during the autumn months.

PANICUM ADSPERSUM, Trin.
"Dense-flowered panick grass."

A perennial species, growing from 1 foot to 1½ feet high, and found only in the arid interior. During the summer months it yields a fair amount of leafy herbage, which stock of all description greedily feed upon, and often so closely that it has little chance to mature any seed. When left unmolested for a time, however, it produces a fair amount of seed, which ripens in October and November.

PANICUM BAILEYI, Benth.
"Bailey's panick grass."

A glabrous perennial species, growing from 1 foot to 3 feet high, and only found, as far as is at present known, in the north-eastern parts of the Colony. It generally grows on good soils, and yields a bulk of leafy herbage, which is much relished by stock of all kinds. If cut when the flower-stems first appear, it makes good hay. It is a grass worth disseminating in the coastal districts, and there should be no difficulty about the matter, as it produces an abundance of seed, which ripens in January and February.

PANICUM BICOLOR, R. Br.
"Two-coloured panick grass."

A perennial, rather slender, tufted grass, rarely exceeding 1½ feet high. It is found growing plentifully in the coastal districts, and more sparingly in New England and south-western parts of the Colony. This grass is not particular as to soil or situation, for it may as often be seen on ironstone ridges as on good pasture land. On good soils, however, it yields a superior herbage, much relished by all herbivora. In sheltered situations, it will yield a fair amount of herbage during the winter and early spring months. This grass produces a fair amount of seed, which ripens in November and December.

PANICUM CŒNICOLUM, F.v.M.
"Finger panick grass."

A perennial species, generally found on chocolate soils in the arid interior, but nowhere very plentiful. The stems rise from a knotty branching base, to about 1½ feet high. During the early summer months this grass yields a fair amount of herbage, much relished by all herbivora. It produces a fair amount of seed, which ripens in October and November.

PANICUM CRUS-GALLI, Linn.
"Barn-yard grass."

An annual species, growing from 2 to 8 feet high, which is generally found in the coastal districts, and in some places fairly plentiful. On moist land this strong grass yields an enormous amount of rich succulent herbage, which is much relished by stock of all kinds. It is especially valuable for milch cows. Some few years ago, I saw this grass cultivated on the low moist lands between Cook's and George's rivers, and bundles of it are sold for green feed in Sydney during the summer months. It is worthy of extensive cultivation on low, moist lands in the coastal districts, not only as supplying valuable forage, but, from the enormous amount of herbage it yields, it ought to make good ensilage. There would be no difficulty in bringing this grass under systematic cultivation, as it produces a great amount of seed, which is easily collected. A collector can easily distinguish this grass by its strongly-bearded panicles. The seeds ripen in February and March. Within the suburban railway line enclosure near Newtown railway station, there was a fine patch of this grass growing last year. This grass is common to all hot and temperate climates. In America it is very highly prized. One writer says "that it gives 5 tons of hay per acre without care or cultivation, and that on the Mississippi hundreds of acres are annually mowed on single farms."

PANICUM DECOMPOSITUM, R. Br.

"Australian millet."

A semi-aquatic, glabrous, annual grass. When found on swampy land it often grows 4 feet high; in drier situations rarely above 1½ feet high. In all its varied forms, however, it yields most valuable forage, which stock of all descriptions are remarkably fond of. Under cultivation, it is a most prolific grass, and if cut when the flower-stems first appear it makes splendid hay. I can highly recommend it for general cultivation. A collector would have no difficulty in gathering any amount of seed of this grass at its season for ripening, which is generally in December and January. The western aborigines used to collect the seed in great quantities, and convert them into cakes. This grass is widely distributed throughout the Colony. Sir Thomas Mitchell (" Three Expeditions," pp. 237 and·290), alluding to this grass, says : " In the neighbourhood of our camp, the grass had been pulled to a very great extent, and piled in hay-ricks, so that the aspect of the desert was softened into the agreeable semblance of a hay-field. The grass had evidently been thus laid up by the natives, but for what purpose we could not imagine. At first, I thought the heaps were only the remains of encampments, as the aborigines sometimes sleep on a little dry grass, but when we found the ricks, or hay-cocks, extending for miles we were quite at a loss to understand why they had been made. All the grass was of one kind, and not a spike of it was left in the soil over the whole ground. . . We were still at a loss to know for what purpose the heaps of one particular kind of grass had been pulled, and so laid up hereabouts. Whether it was accumulated by the natives to allure birds, or by rats, as their holes were seen beneath, we were puzzled to determine. The grass was beautifully green beneath the heaps, and full of seeds, and our cattle were very fond of this hay."

PANICUM DISTACHYUM, Linn.

" Two-spiked panick grass."

A slender perennial species, with decumbent or creeping stems, rooting at the lower nodes, and ascending to about 1⅓ feet. It is mostly found in the warmer parts of the Colony, and in some situations it is fairly plentiful. It is a very superior pasture grass, and during the summer months yields a large quantity of valuable herbage, which stock are very fond of.· If cut when the flower stems first appear it makes excellent hay. Under cultivation this is a most productive grass, and I can highly recommend it, either for permanent pasture or hay. When left unmolested for a time it produces a fair quantity of seed, which ripens in November and December.

PANICUM DIVARICATISSIMUM, R. Br.

" Umbrella grass."

A perennial species growing from 1 foot to 2 feet high, and generally found from the coast to the arid interior. There are several varieties of this excellent pasture grass, but their qualities are much the same. Although its large spreading panicles give it an uninviting appearance, still it yields a lot of valuable herbage during the summer months, which stock of all kinds are remarkably fond of. The roots of this grass are tough and fibrous, and they penetrate deeply into the soil, which enables the plant to withstand the most prolonged drought. This grass produces a great amount of seed, which ripens in October and November.

PANICUM EFFUSUM, R. Br.
"Branched panick grass."

An erect perennial species, growing from 1 foot to 2 feet high, and generally found all over the Colony. It is not particular as to soil or situation, for it may be seen on dry ironstone ridges as often as on the more fertile low-lands. In the latter situation, however, it yields a rich succulent herbage, which is much relished by all herbivora. The whole plant is hairy, but there is a variety (var. *convallium*) growing in the western districts which is quite glabrous ; with this exception, however, its qualities are much the same. Under ordinary circumstances this grass yields a fair quantity of seed, which ripens in October and November.

PANICUM FLAVIDUM, Retz.
"Yellow-flowered panick grass."

An erect perennial grass, growing from 1 foot to 3 feet high, and generally found all over the arid interior. On good soils it produces a great quantity of rich succulent herbage, which stock of all kinds are particularly fond of. This grass is held in much esteem with pastoralists, and it is said to be one of the best fattening grasses in the interior. The tough fibrous roots of this grass penetrate deeply into the soil, which enables the plant to withstand the most protracted drought. When brought under cultivation, this grass yields an enormous amount of herbage, and if cut when the flower-stalks first appear, it makes splendid hay. This grass would well repay systematic cultivation on our arid plains. After a crop of hay was taken off it would make good pasture for the rest of the summer. There should be no difficulty in bringing this grass under cultivation, as it produces an enormous amount of seed—in fact, the seed stalks are so heavily loaded with grain that they often lie prostrate on the ground. The seeds ripen in October and November. There is a variety (var. *tenior*) of this grass which generally grows in the coastal districts, and although it rarely ever exceeds 1 foot in height it is a good pasture grass, and in sheltered situations will grow nearly all the year round. It is a capital grass to withstand dry weather, and its broad green leaves may often be seen when the surrounding grasses are dried up. In some situations about Port Jackson it is quite common. It is a prolific seed-bearer, and the stems are often prostrate from the weight of grain.

PANICUM FOLIOSUM, R. Br.
"Leafy panick grass."

A rather handsome, broad-leaved, hairy grass, rarely exceeding 2 feet high, and generally found on river banks and on the borders of scrubs in the north-eastern parts of the Colony, but nowhere very plentiful. It is not a good pasture grass, as its roots have so slight a hold of the soil that stock often destroy it when browsing upon its herbage, by pulling it out of the ground. This grass produces a fair amount of seed, which ripens in December and January.

PANICUM GRACILE, R. Br.
" Slender panick grass."

An erect many-branched perennial grass, rarely exceeding 1½ feet in height, and generally found all over the Colony. It is an exceedingly variable grass as regards stature and appearance, and some forms of it might readily be mistaken for the variety *tenior* of *P. flavidum.* This grass is not particular as to soil or situation, and it may as often be found on hill sides as in the more fertile pastures. Although its leaves are narrow and somewhat harsh in dry seasons, it is, nevertheless, a good pasture grass, and one which stock of all descriptions are fond of. This grass does not appear to be a great seed bearer : what there is, however, ripens in October and November in the interior, and December and January in the coastal districts.

PANICUM HELOPUS, Trin.
" Hairy panick grass."

An erect and somewhat hairy perennial grass, growing from 1 foot to 2½ feet high, and only found in the arid interior, but in some places it is fairly plentiful. It generally grows on good soils, and during the early summer months yields a rich, succulent, herbage much relished by all herbivora. This grass produces a fair amount of seed, which ripens in October and November.

PANICUM INDICUM, Linn.
" Indian panick grass."

An erect, slender, wiry, perennial grass, growing from 8 inches to 1½ feet high, and generally found in low, damp, localities in the coastal districts. It is of little value as a pasture grass, and, unless in very dry seasons, when other herbage is scarce, stock seldom or never touch it. It produces a fair amount of seed, which ripens in December and January.

PANICUM LEUCOPHŒUM, H.B. et K.

An erect perennial grass, growing from 1 foot to 3 feet high, and is generally found over a greater portion of the interior. It is a valuable pasture grass, and during the summer months yields a quantity of valuable herbage, which is much relished by stock of all kinds. Under cultivation this is a most prolific grass, and if cut when the flower stalks first appear it makes excellent hay. I can highly recommend this grass for general pasture, or for making into hay. This grass is easily recognised in pastures by its spikelets being densely covered with long, silky, silvery, or purple hairs, which gives it quite an ornamental appearance. It produces a fair amount of seed, which ripens in November and December. There is a variety of this grass (var. *monostachyum*), which is more dwarf in habit, and has the inflorescence arranged in a simple spike instead of a panicle. With these exceptions, however, its qualities are much the same. This variety is generally found on ridges in the interior. It produces a fair amount of seed, which ripens in November.

PANICUM MACRACTINIUM, Benth.
" Roly-poly grass," " Umbrella grass."

A perennial grass, of somewhat tufted habit, and generally found all over the arid interior. Its immense spreading panicles give it an uninviting appearance. Notwithstanding this, however, it is an excellent pasture grass, and forms a good tuft of leafy herbage at the bottom, which stock are fond of. In the western districts I have seen this grass beautifully green in January when the surrounding herbage was dry looking. But this may be accounted for by its tough, fibrous roots, penetrating deeply into the soil and helping it to sustain its verdure under trying circumstances. Under cultivation this grass is much improved as regards the bulk of herbage it will yield, and if cut when the flower stalks first appear it can be made into excellent hay. I can highly recommend this grass for permanent pasture. It produces a fair amount of seed, which ripens in November and December.

PANICUM MARGINATUM, R. Br.

"Variable panick grass."

A rather slender, rigid, perennial grass, with decumbent stems, and generally found in the coastal districts. It is a most variable grass, both as regards the size of the stems, leaves, and inflorescence, and often a difficulty is experienced in uniting all its varied forms under one species. The only way that this can be done accurately is by examination of the fruiting glume, which is always densely hairy, a peculiarity that has not been observed in any other Panicum. This grass is of no value for forage, although it might be useful for binding embankments and any loose earth. This grass produces a fair amount of seed, which ripens throughout the summer months.

PANICUM MELANANTHUM, F.v.M.

"Black panick grass."

An annual (?) glabrous grass, growing from 1 foot to 3 feet high. It is generally found in damp places in the coastal districts, in New England, and in some of the western districts, but it is not plentiful anywhere. It is worthy of dissemination, however, on all moist lands, for in such situations, during the summer months, it produces a fair amount of herbage, which is much relished by stock. This species is easily distinguished by its large panicles of dark coloured seeds. It produces a fair amount of seed, which ripens in December and January.

PANICUM MITCHELLII, Benth.

"Mitchell's panick grass."

An erect perennial grass, growing from 2 to 4 feet high, and is generally found in the arid interior, but also more sparingly in the north-eastern districts. On the rich loamy plains of the interior it produces a great amount of rich, succulent herbage, which stock of all descriptions are remarkably fond of. Under cultivation it is a most prolific grass, and if cut directly the flower stems appear it makes splendid hay. I can highly recommend this grass, either for general pasture or to be grown for hay, and it is even bulky enough to be used for ensilage. It produces an abundance of seed, which ripens in November and December.

PANICUM OBSEPTUM, Trin.

"Mud panick grass."

A weak glabrous perennial grass, with the bases of the stems creeping in the mud and shortly ascending. It is usually found on the borders of ponds, both in the coastal and New England districts, but nowhere very plentiful. It is of very little value for forage, but cattle will occasionally browse on it. This grass produces very little seed: what little there is, however, ripens during the summer months.

D

PANICUM PARVIFLORUM, R. Br.

" Small-flowered panick grass."

A slender glabrous perennial grass, growing from 1 foot to 3 feet high, and generally found in the coastal districts, and in some places it is fairly plentiful. In sheltered situations it will grow all the year round, but during the summer months it yields a quantity of rich succulent herbage, which is greedily eaten by all herbivora. This grass will remain green during a long spell of dry weather, and can be highly recommended as a permanent pasture grass, or for making into hay. It produces an abundance of seed, which ripens in December and January. There is a variety (var, *pilosa*) of this grass but with the exception of being hairy it has all the characteristics of the species.

PANICUM PROLUTUM, F.v.M.

" Rigid panick grass."

An erect, rather rigid, perennial species, growing from 1 foot to 2½ feet high, and principally found in the interior, where, however, it is moderately plentiful in some situations. It generally grows on good land that is liable to periodical inundations, and as it makes most of its growth during the summer months, it is a valuable stand-by for stock, when many other grasses are somewhat scarce. It is a valuable grass for withstanding a long spell of dry weather, and, under ordinary circumstances, will retain its greenness far into the autumn months. It is not a good grass to make hay of, as its stems and leaves are too rigid. Before the aborigines tasted the sweets of civilisation, they used to collect the seeds of this grass in large quantities, and use them as an article of food, after being ground between two stones and converted into a kind of meal. This grass produces an abundance of seed which ripens at various times of the year.

PANICUM PYGMÆUM, R. Br.

" Pigmy panick grass."

A small hairy creeping perennial grass, principally found in shady places in the coastal districts. It is of no value as a forage grass, for stock seldom or never touch it; but it is useful for covering ground under the shade of trees. It forms a beautiful green sward in a short time, and might be utilised in gardens for covering shady places where scarcely anything else would grow. It can easily be propagated by division of its stems, and also from seed, which it bears in fair quantities. The seeds ripen during the autumn months.

PANICUM REPENS, Linn.

" Creeping panick grass."

A perennial species, with a creeping and rooting base, from which stems rise to 2½ feet high. It is rather a rare grass in New South Wales, and I have found it only on the Murrumbidgee River. It delights to grow in a moist rich soil, and it is worth disseminating on the banks of rivers, dams, &c, which its strong roots would help to bind, and protect against the fury of flood waters. It is a rather stiff growing grass, and stock only eat it when other herbage becomes scarce. It is a good seed-bearing grass. The seeds ripen in January and February. This grass can be propagated by division of its roots.

PANICUM REVERSUM, F.v.M.

"Reversed panick grass."

A perennial grass, with many-branched but weak stems, and long narrow glaucous leaves. It is found only in the arid interior, but nowhere plentifully. Where it does occur, however, stock will browse upon it. This grass produces very little seed: what there is ripens in November and December.

PANICUM SANGUINALE, Linn.

"Summer grass."

An annual species, which is common all over the eastern portion of the Colony. It is a creeping quick-growing grass, and a great pest in cultivated ground to farmers, orchardists, and gardeners. It will grow in almost any kind of soil, and in any situation, provided it is not too cold. This grass produces a great amount of forage in an incredible short space of time, and being of a succulent nature, is relished by all pasture animals. In America this grass is highly spoken of, and it is said that horses are so fond of the hay made from it that they leave all other fodder for it. This grass produces an abundance of seed, which ripens in January, February, and March. It is said that Linnæus gave the specific name "Sanguinale," from a trick that the boys had in Germany of pricking one another's noses with the spikes of this grass until they bled.

The *Hortus Gramineus Woburnensis* says of this grass:—" It produces much seed, of which birds are very fond, and requires to be protected by nets, or otherwise, during the time of ripening. The smaller birds pick out the ripe seed, even when only a small quantity is formed among the blossoms. The common method of collecting it and preparing it in Germany is as follows:—At sunrise the grass is gathered or beaten into a hair sieve from the dewy grass, spread on a sheet, and dried for a fortnight in the sun; it is then gently beaten with a wooden pestle in a wooden trough or mortar, with straw laid between the seeds and the pestle till the chaff comes off; they are then winnowed. After this they are again put into a trough or mortar in rows, with dried marigold flowers, apple, and hazel leaves, and pounded till they appear bright; they are then winnowed again, and being made perfectly clean by this last process are fit for use. The marigold leaves are added to give the seed a finer colour. A bushel of seed with the chaff yields only about two quarts of clean seed. When boiled with milk and wine it forms an extremely palatable food, and is in general made use of whole in the manner of sago, to which it is in most instances preferred."

PANICUM SEMIALATUM, R. Br.

"Cockatoo grass."

An erect perennial grass, growing from 2 to 3 or more feet high, which is found both in the north-eastern and north-western districts, but nowhere very plentifully as far as I have been able to make out. It is a valuable drought-resisting grass, and an excellent one for pasture. During the summer months this grass yields a large amount of leafy forage which stock are fond of. In the autumn months, when it is ripening its seeds, the stems become hard and cane like; then stock seldom or never touch it. This grass is worth disseminating in our coastal districts, for once it gets fairly established in the soil it takes a lot of dry weather to kill it. It will grow in almost any kind of soil, but on good pasture land it would yield an enormous amount of herbage which might bo turned into ensilage. Besides its valuable forage, this grass perfects a fair sized grain, which cockatoos are said to be extremely fond of It is a prolific seed-bearing grass when left undisturbed for a time. The seeds ripen during the autumn months.

PANICUM TENUISSIMUM, Benth.
" Very slender panick grass."

A very slender perennial grass, growing from 1 foot to 2 feet high, which is found in the north-eastern parts of the Colony, but not very plentifully. During the summer months it yields a fair amount of herbage which is relished by stock. It does not produce much seed : what little there is, however, ripens in December.

PANICUM TRACHYRHACHIS, Benth.
" Coolibar grass."

A stout glabrous perennial grass, growing from 2 to 3 feet high, which is principally found in the north-western interior, and in some places, is abundant. It generally grows on rich soils on open downs country, and during the summer months yields a great amount of valuable herbage which stock of all descriptions are fond of. This grass would well repay systematic cultivation, either for general pasture or for hay. It is a prolific seed-bearing grass, and one of those from which the blacks at one time gathered a great amount of grain and used it largely as an article of food, after grinding it between stones and making it into a kind of meal. The seeds ripen in October and November.

PANICUM UNCINULATUM, R. Br.
" Hooked panick grass."

A coarse, erect, many-branched, perennial grass, sometimes attaining a height of 8 feet. It is generally found growing on mountain sides, and in scrubby country under the shade of trees, both in the coastal districts and north-western interior, and in some places it is fairly plentiful. This coarse grass is of little value for forage, except after being burnt off ; and if this should take place during October or November, it will produce a fair amount of succulent herbage for a greater part of the summer, which stock will graze upon. This grass does not produce much seed : what little there is, however, ripens in January and February.

PAPPOPHORUM AVENACEUM, Lindl.
" White heads."

A perennial grass, rarely exceeding, 1¼ feet in height, which is generally found on rich soils in the interior. It is a first-class drought-resisting species, and during the summer months yields a fair amount of good herbage, which is much liked by all herbivora. This grass is worthy of being widely disseminated on our arid plains, for, in the most trying seasons, it can be depended upon to supply some herbage. It produces a great amount of seed, which ripens in October and November.

PAPPOPHORUM NIGRICANS, R. Br.
"Black heads."

An erect perennial species, rarely exceeding 2 feet in height. It is generally found growing all over the Colony, from the coast to the arid interior. As might be supposed, a grass growing under such varied conditions of soil and climate is most variable in habit, and also with regard to the colour of its inflorescence. Sometimes it is perfectly black, which circumstance led to the specific name *nigricans* being given to it; at other times it is almost white, but the grass can never be mistaken under microscopical examination. It is a capital drought-resisting species, and during the early summer months yields a fair amount of good herbage, which stock are fond of and fatten on. When this grass becomes old, however, the stems get rather hard and wiry, and if other grasses are plentiful, stock will seldom touch it. Some of the forms of this grass are very ornamental, and are worthy of the attention of horticulturists. This grass produces an abundance of seed, which ripens in October, November, and December.

PASPALUM BREVIFOLIUM, Flügge.
" Short-leaved paspalum."

A slender, tufted, perennial grass, with creeping underground stems from which spring short broad leaves, which is widely spread in the coastal districts. The short tender herbage is good forage for sheep, and it will stand close feeding. It is an excellent grass for a lawn, as it takes so very little trouble to keep in order. This I have had experience of; and, besides, it will keep its verdure throughout the summer, when grown on fairly good soils. Its delicate flower stems generally grow from 1 foot to 2 feet high, but it does not produce a great amount of seed; what little there is, however, ripens in January and February. This grass is easily propagated by division of its roots.

PASPALUM DISTICHUM, Linn.
" Water-couch."

A perennial grass, with creeping, rapid-growing, succulent stems, generally growing in swampy places, but sometimes in water, and always in the coastal districts. It yields a great quantity of valuable herbage, of which stock of all descriptions are remarkably fond. Butter made from the milk of cows fed exclusively on this grass is quite white, but in no other way is it affected. It is a poor grass, however, for making into hay, as it turns black in drying. This grass is exceptionally well adapted for covering waste moist lands, the banks of rivers, and dams, which it binds very firmly once its underground stems get well established. Periodical inundations will not destroy it, but it is injured by frosts. It remains beautifully green throughout the summer months, and some persons have been tempted to plant it on lawns, with rather serious consequences, however, for to keep it in anything like order during the summer months it requires cutting two or three times a week, and it is as bad as ordinary couch to get out of cultivated land. This grass produces an abundance of seed, which ripens in January, February, and March. There is a variety of this grass (var. *littorale*), which is only found in or near brackish swamps, and only differs from the one described last by its narrower leaves. With these exceptions, its qualities are much the same.

PASPALUM MINUTIFLORUM, Steud.
" Small-flowered paspalum."

An erect glabrous grass, growing from 1 foot to 2½ feet high, and only found in the north-eastern part of the Colony, and generally on damp soil. During the summer months this grass yields a lot of valuable herbage, which cattle greedily feed upon. It will also make good hay if it is cut before the flower stems become too old. This grass produces an abundance of seed, which ripens in January, February, and March.

PASPALUM SCROBICULATUM, Linn.
" Ditch millet."

An erect, quick-growing perennial grass, generally found in wet land in the coastal districts, where it usually attains a height of 1 foot or 2 feet. I have found this grass growing quite common near Rookwood. It is of little or no value for pasture, and, except when it makes its young growth in spring, stock seldom or never touch it. During the autumn months the flowers of this grass are terribly subject to parasitic fungi. Fifteen years ago I first observed fungoid growth on this grass, and it occurs annually, often blighting the whole panicle. What seed this grass does produce it ripens in March and April. In Lindley's Vegetable Kingdom, p. 113, speaking of injurious grasses, he says :—" And a variety of *Paspalum scrobiculatum*, called Hureek in India, which is perhaps the Ghohona grass, a reputated Indian poisonous species, said to render the milk of cows that graze upon it narcotic and drastic. . . . . The Monya or Kodro of India, a cheap grain, regarded as wholesome, is another variety of this species."   Church " Food Grains of India," gives the following composition of Kodro or Koda Millet (husked) :—

						In 100 parts.	In 1 lb. oz. gr.
Water,	...	...	...	...	...	11·7	1·382
Albuminoids...	...	...	...	...·	7·0	1·52
Starch	...	...	...	...	...	77·2	12·154
Oil	...	...	...	...	...	2·1	0·147
Fibre	...	...	...	...	...	0·7	0·49
Ash ...	...	...	...	...	...	1·3	0·91

PENNISETUM COMPRESSUM, R. Br.
" Swamp fox-tail grass."

A very scabrous perennial species, growing into large tussocks, and generally found in or about swamps, in the coastal districts, but also in New England. To render this grass of any value for forage it requires to be burnt off annually ; then for a few months the herbage, although coarse, is eaten by stock. This grass has on more than one occasion proved a valuable stand-by for stock when drought has been stalking throughout the land and other herbage burnt up. It produces a fair amount of seed, which ripens in January and February.

PENTAPOGON BILLARDIERI, R. Br.
" Five-awned grass.

A hairy-leaved annual grass, growing from 1 foot to 2 feet high, and only found in the southern districts. It is not considered a good pasture grass, although sheep will eat it, but in doing so they sometimes pull the plant out of the ground. This grass produces a fair amount of seed, which ripens in February and March.

PEROTIS RARA, R. Br.
" Comet grass."

A slender growing perennial grass, rarely exceeding a foot in height, and principally found in the north-western interior. It generally grows on rich soils in open country, and during the early summer months, whilst the herbage is green, is freely eaten by sheep. When it becomes old, however, it is seldom eaten, for it has rather a forbidding appearance. Its flower spikes are often 8 inches long, and the spikelets taper into long terminal straight awns, which are often 1 inch long. It is said, however, that these awns are troublesome neither to the sheep, nor to their wool. This grass produces a fair amount of seed which ripens in October and November.

PHRAGMITES COMMUNIS, Trin.
"The common reed."

A very stout perennial grass, sometimes only 5 feet, at other times 12 feet high, and is very abundant on the margins of rivers and in swamps. It is not of much value from an agricultural point of view, for unless in very dry seasons stock seldom eat it. It is a grass of much importance, however, for binding the banks of rivers which are subject to periodical floods. Once its underground extensively creeping stems get well established in the soil, scarcely anything can move them. It is most easily propagated by division of its roots. It also produces a great quantity of seed which ripens at various times of the year. The aborigines at one time made their baskets from the stems of this reed, as also their light spear-handles. This reed when dry yields 4·7 per cent. of ash, which, according to Schulz-Fleeth (Watts, Dict., i., 413), contains in 100 parts:—

Potash (anhydrous)	8·6
Lime...	5·9
Magnesia	1·2
Ferric oxide...	0·2
Sulphuric acid (anhydride)	2·8
Silica	71·5
Carbonic acid	6·6
Phosphoric acid ($P_2 O_5$)	2·0
Sodium chloride (common salt)	0·4

POA CŒSPITOSA, Forst.
"Tussock poa."

A perennial species, growing from 1 foot to 3½ feet high, and generally found all over the Colony. Abundant in the coastal districts, but more sparingly distributed over the interior. It is an exceedingly variable grass. Besides the typical form, there are five well defined varieties, and, as might be supposed, they vary considerably in the amount of herbage each one yields. All of them, however, are excellent pasture grasses, and stock of all kinds are remarkably fond of them. They are capital drought-resisting grasses, and if not allowed to go to seed, will grow and remain green during a greater part of summer. Nearly all of them produce an abundance of seed, which ripens from November to March. There is a tall and luxuriant variety (var. *latifolia*), with leaves over a quarter of an inch broad, found in the Illawarra District, and on the Muniong Mountains that is well worthy of extensive cultivation. Besides yielding a large amount of rich succulent herbage, it will, if cut before the flower-stems appear, make excellent hay.

POA LEPIDA, F.v.M.
"Scaly poa."

An erect annual grass, growing from 2 to 3 inches to nearly a foot high, and only found in the interior. It very much resembles that ubiquitous introduced annual, *Poa annua*, and like that species is of very little value for forage. This grass does not produce much seed: what little there is, however, ripens in September and October.

POA NODOSA, Nees.
"Nodding poa."

A perennial (?) species, usually attaining a height of 2 feet, and only found in the southern parts of the Colony. This is the most distinct of all the Australian Poas. Besides its broad spikelets, which are almost like those of the large quaking grass (Briza), there are two or three globular or ovoid nodules at the base of each stem. I have observed this in very small specimens. During the summer months this grass produces a fair amount of rich succulent herbage, which sheep are fond of. It is worth disseminating in the colder parts of the Colony. This grass produces a fair amount of seed, which ripens in January and February.

POLLINIA FULVA, Benth.
"Sugar grass."

A tall perennial species, growing from 2 to 4 or even more feet high, which is generally found on moist land, and along water courses in the interior, but nowhere very plentifully. This grass is easily recognised by its rich brown silky spikes of flowers. It is a superior pasture grass, and during the summer months it produces a great amount of rich succulent sweet herbage, which is much relished by all herbivora, and, if cut before the flower stems appear, it makes excellent hay. Under cultivation, this grass produces an amount of forage that is quite astonishing, and I can highly recommend it for permanent pasture or hay, or it is even bulky enough to make into ensilage. This grass is much praised by stockowners, and they have given it the common name of "sugar grass," on account of the sweetness of its stems and foliage. When left unmolested for a time, it produces a fair amount of seed, which ripens in November and December.

POTAMOPHILA PARVIFLORA, R. Br.
"Hastings River reed."

An aquatic glabrous perennial grass of 3 to 5 feet, which is found only in the north-eastern parts of the Colony, but withal so rare, that little is known of its qualities as a forage plant. This grass bears panicles of flowers between 1 foot and 2 feet long, of a pale silvery or purple colour, which gives it a striking appearance. The seeds ripen in February and March.

SCHEDONORUS HOOKERIANUS, Benth.
"Hooker's fescue grass."

A stout, glabrous, perennial grass, growing from 2 to 4 feet high, which is principally found in the southern parts of the Colony, where, however, it is not very plentiful. It is a good pasture grass, and worthy of being widely disseminated in all the colder parts of the Colony. Its herbage is greedily eaten by all herbivora. Its loose panicles are often a foot long. This grass produces a fair amount of seed, which ripens in February and March.

SCHEDONORUS LITTORALIS, Beauv.
"Coast fescue grass."

A perennial species, forming dense hard tufts of a pale yellow colour. The stems rise from 1 foot to 3 or more feet high. It is always found on the littoral sands, and is of much importance in such situations, and it might be extensively used for binding drift-sand on our shores. There are one or two varieties of this useful grass, but their qualities are much the same, their only difference being that some of them are smaller in size, both in stems and inflorescence. I have found the typical form of this grass on the beach at Bondi; but it is of little or no value as a forage plant, for the leaves and stems are so thickly sprinkled with sand as to render them unfit for food. It should be mentioned, however, that all littoral grasses contain a considerable amount of soda in their stems and leaves, which is invaluable to the health of stock. These grasses can be propagated by division of their roots. They bear a fair amount of seed which ripens in the summer months.

SETARIA GLAUCA, Beauv.
" Pigeon grass."

An erect annual, of a pale green colour, and growing from 2 to 3½ feet high. It is generally found all over the Colony, but not in all places plentifully. On rich soils, or on land that has been newly broken up, it yields a rich succulent herbage during the summer months, which is much relished by stock of all kinds. This grass is worthy of systematic cultivation, either to be cut for green feed, or for grazing, or for making into hay. If for the last, it should be cut when the flower stems first appear. Under cultivation it yields a surprising amount of forage, which might be turned into ensilage. This grass produces an abundance of seed, which might be put to some economic use. The seeds ripen during the autumn months.

SETARIA MACROSTACHYA, H. B. et K.
" Large-headed setaria."

A very coarse-growing annual grass, sometimes attaining 6 feet in height, and only found in the north-eastern parts of the Colony. It is generally found growing on rich scrub lands, and often in cultivation, where it yields a rich succulent herbage, which cattle are particularly fond of. This grass would well repay systematic cultivation, especially where dairy cows are kept, and it would likely make good ensilage. It is a prolific seed-bearing grass, and the seeds might be put to some economic use. They ripen during the autumn months.

SORGHUM HALEPENSE, Pers.

This grass has several common names, which often are confounding, such as the Cuba-grass, Johnson-grass, and Haleppo-grass. The last, however, is the original appellation, and should be retained. An erect perennial grass, often growing 10 feet high on good soils, and now is found in various parts of the Colony. Its value for forage has long been recognised, and seeds of it are now obtainable in the trade. Although this grass will grow on various soils, and in different situations, still, to get satisfactory returns, it should always be sown on rich deep soils, where its strong fibrous roots can penetrate deeply into the earth. In these circumstances the grass will remain green during the driest of weather. Under ordinary circumstances this grass will stand cutting three or four times during one season. It is a prolific seed-bearer, and the seeds ripen during the autumn months.

SORGHUM PLUMOSUM, Beauv.
" Plumose sorghum."

A strong-growing perennial grass, attaining a height of from 4 to 8 feet, and generally found in the coastal districts, but I have had specimens also from

New England. It is almost as valuable a pasture grass as the last one, but unfortunately too little known to the cultivator. It is a good drought-resisting grass, and during the driest of seasons will yield a great quantity of valuable herbage, which is much relished by all herbivora. This grass is well worthy of systematic cultivation, both in the coastal districts and colder parts of the Colony, where it could not only be grown for forage and hay, but from its bulky yield would make good ensilage. It is a prolific seed-bearer, and the seeds ripen during the autumn months.

SPINIFEX HIRSUTUS, Labill.
"Spiny rolling-grass."

A perennial species, with stout creeping stems which root at every joint; it often forms large tufts on the littoral sands. This grass is diœcious—that is, the male and the female inflorescence are borne on separate plants. Its very peculiar inflorescence is often gathered for ornamental purposes. At one time this grass was abundant at Lady Robinson's Beach, and is still growing plentifully on the sandy shores of Botany Bay and at many other places on the coast of New South Wales. It is of no value as a forage grass, as stock seldom or never touch it, but it is most useful for fixing drift sands when encroaching on valuable land. For this purpose it deserves more attention than has hitherto been paid to it. Some years ago I highly recommended this grass to be planted on the drift sands at Wollongong and Newcastle. It is of comparatively quick growth, and once it gets well established on the sand scarcely anything will kill it; even the spray from the salt-water will not check its growth. This grass is very easily propagated by pieces of the stems. August and September are the best months for doing this.

SPINIFEX PARADOXUS, Benth.
"Curious spinifex."

Another diœcious grass, but not so stout as the preceding species, and, besides, it is only found in the arid interior, where it generally grows on sand-ridges, which it often covers to the exclusion of other grasses. The drought-enduring qualities of this grass are something remarkable, for it is neither affected by the fierce heat of the summer's sun, nor by hot winds that periodically blow in the summer months on our western plains. When other herbage is scarce, it affords a quantity of valuable forage for stock. This grass produces a fair amount of seed, which ripens in November and December.

SPOROBOLUS ACTINOCLADUS, F.v.M.
"Whorled sporobolus."

A rather delicate, perennial grass, growing from 1 foot to 2 feet high, and only found on the arid plains in the interior. During the spring and early summer months it yields a fair amount of succulent herbage, which is much relished by all herbivora. This grass produces plenty of seed, which ripens in October and November.

SPOROBOLUS DIANDER, Beauv.
"Tussock grass."

A glabrous perennial species, growing from 1 foot to 3 feet high, and only found in the north-eastern portion of the Colony. On rich soils, and on land that is liable to periodical inundations, it forms good-sized tussocks of deep green herbage, and whilst it is young is much relished by stock. When it becomes old, however, the herbage is very tough ; then cattle seldom or never touch it. This grass might be utilized for paper-making. It produces plenty of seed, which ripens during the autumn months.

SPOROBOLUS INDICUS, R. Br.

"Parramatta or tussock grass."

An erect, tufted, perennial grass, of 1 foot to 2½ feet, and generally found in the coastal districts, and in some places very abundant. In fact, in some places, where land has been broken up and sown with exotic grasses, the Parramatta grass is now master of the situation, much to the disgust of dairy farmers. Whilst young, it affords capital feed, but when old is very tough and wiry, so much so that it will loosen the teeth of horses and cows when kept too long on pasture where this grass predominates. I have often recommended this species for paper-making. It seems to be as strong as the esparto grass (*Stipa tenacissima*) of Spain when they are grown side by side in Australia. This grass is a prolific seed-bearer, and the seeds are eaten by many small birds. They ripen at various times of the year. There is a variety of this grass (var. *elongatus*) with narrower leaves, and a longer and looser panicle. With these exceptions, however, its qualities are much the same.

SPOROBOLUS LINDLEYI, Benth.

"Lindley's sporobolus."

An exceedingly pretty perennial grass, growing about 1 foot high, and only found on rich soils in the interior, but nowhere very abundant. During the winter and early spring months it yields a fair quantity of tender herbage, of which sheep are remarkably fond. At one time the seeds of this grass were collected and used as an article of food by the aborigines. It is a prolific seed-bearer. The seeds ripen in October and November.

SPOROBOLUS PULCHELLUS, R. Br.

"Pretty sporobolus."

Another very pretty perennial grass, growing from 6 inches to a foot high, which is principally found on land liable to periodical inundations in the interior. It is, however, a rather rare species in this Colony. This grass makes its growth during the winter and early spring months, and its tender herbage is eaten by sheep. It does not produce much seed ; what little there is, however, ripens in October and November.

SPOROBOLUS VIRGINICUS, Kunth.

"Salt marsh couch grass."

A perennial species, with underground creeping stems, from which spring a quantity of leafy herbage, but rarely exceeding a foot in height. This grass is found on the salt marshes in the coastal districts, and in some places it is fairly plentiful. During a greater part of the year it affords a splendid herbage, which stock are remarkably fond of, and they fatten well on it. This littoral grass contains a considerable amount of soda in its stems and leaves, which is invaluable to the health of stock. It is well worthy of cultivation on any salt marshes where it may not be growing, as it could always be relied upon to yield a good herbage during long spells of dry weather, when other grasses are scarce. This grass is easily propagated by division of its underground stems. The spring of the year is the best time for this operation. It produces a fair amount of seed, which ripens in November and December. There is a variety of this grass (var. *pallida*) which is found on the Richmond and Darling Rivers. (In the latter place it grows on loose sand but it is not plentiful.) Except that it is a little taller and narrower in the leaf, and has a looser flower-spike, its qualities are much the same.

STIPA ARISTIGLUMIS, F.v.M.
"Spear-grass."

A perennial species, growing from 2 to 3 or more feet high, and is found on rich soils, both in the coastal districts and north-western interior, and in some situations it is fairly plentiful. After rains in springtime it makes a wonderful quick growth, which horses, cattle, and sheep are remarkably fond of, and graziers in general consider it a very fattening grass. When the herbage becomes old, however, it is coarse and rather harsh, and unless other grasses are scarce, stock will seldom or never touch it. This grass should be burnt off annually, for when this is done the herbage is much improved for a time, and, besides, the troublesome seed awns are destroyed. These, however, are not so troublesome in this as in some of the allied species. This grass produces an abundance of seed, which ripens in November and December.

STIPA ELEGANTISSIMA, Labill.
"Silver-plumed spear grass."

A rather slender many-branched perennial grass, growing from 2 to 4 feet high, and only found in the interior. It is the most elegant of all the Stipas, which is saying a great deal, and is as well worthy of cultivation as the allied one (*Stipa pennata*) which may often be seen in gardens. This grass is usually found growing beneath the shelter of some thick bush, 3 or 4 feet high. At the flowering season, the elegant plumose panicles force their way through the bush and cover the whole with a mass of beautiful silver network, forming a conspicuous object. This grass is only occasionally eaten by stock. The seed awns are not injurious to sheep or cattle. Its seeds ripen in October and November.

STIPA FLAVESCENS, Labill.
"Spear grass."

A rather rigid perennial grass, growing from 1½ to 3 feet high, and is only found on the Maneroo Plains and in the extreme southern parts of the Colony. It is of little value, however, as a forage plant, as its fine short leaves soon become hard and wiry. It is one of the troublesome spear-grasses which could be very well spared from pastures. The seed awns of this grass are about 1½ inches long, and often are irritating to the eyes of sheep. Its seeds ripen in November and December.

STIPA MICRANTHA, Cav.
"Bamboo spear grass."

A perennial species, with rigid spreading branches, often several feet high. This grass is found in the coastal districts and also in New England, and on low moist lands it is fairly plentiful. It sprouts prolifically from the joints, and during the early part of the year these dense tufts of leaves are much browsed upon by cattle and horses; the latter seem to be particularly fond of this grass. It is much improved by being burnt off annually. When this is done in the spring of the year, it will make a lot of valuable herbage during a greater part of the summer. The seed awns of this grass are very small; consequently they are not at all troublesome to stock. The seeds ripen in December, January, and February.

STIPA PUBESCENS, R. Br.

" Spear grass."

A tall perennial species, often over 3 feet high, and generally abundant in the coastal districts, but more sparingly on the Blue Mountains and in New England. This grass produces a fair amount of herbage, and whilst young is much eaten by stock. When it becomes old, however, the herbage is hard and wiry, and is seldom or never eaten. This grass should be burnt off annually. Its seed awns, which are often over 2 inches long, are very troublesome to the eyes of sheep, and often get entangled in and deteriorate the wool. The seeds ripen in November and December.

STIPA SCABRA, Lindl.

" Spear grass."

A rather slender tufted perennial grass, rarely exceeding 2 feet high, which is found only on dry soils in the arid interior. It is a capital drought-resisting species, and during the summer months it yields a fair amount of herbage, of which sheep are particularly fond. When the seeds of this grass are ripe, however, their adherent awns, which are often more than 3 inches long, are very troublesome to the eyes of sheep, and are often difficult to get out of the wool. This grass does not produce so much seed, however, as some of its congeners. What there is, however, ripens in November and December.

STIPA SEMIBARBATA, R. Br.

" Spear grass."

A stout perennial species, growing from 2 to 3 or more feet high, which is generally found in the colder parts of the Colony. During the early part of summer it yields a fair amount of herbage, which cattle and horses graze upon. In the autumn months, however, the herbage becomes very hard and wiry; then stock seldom touch it. This grass should be burnt off annually, for besides destroying this coarse herbage, most of its dangerous seeds and seed awns would be consumed. The panicles of this grass are often 10 inches long, and they bear numerous seeds, the awns of which are sometimes nearly 4 inches long. It is an undesirable grass to have in pastures, for the seed awns are not only dangerous to the eyes of sheep, but they get entangled in the wool. The seeds ripen in November, December, and January. There is a variety of this grass (var. *mollis*) which is somewhat coarser, and the foliage is usually covered with soft hairs. With these exceptions, however, its qualities are much the same. This variety is mostly found in the coastal districts.

STIPA SETACEA.

"Spear grass."   "Corkscrew grass."

A rather coarse perennial species, growing from 1 foot to nearly 3 feet high, which is generally found on good soils all over the Colony, and in some places very abundant.   It is an excellent pasture grass, whilst the herbage is young ; but, like several other of its congeners, the herbage gets too harsh and wiry when old.   This grass should be burnt off annually, which destroys both the coarse foliage and the dangerous seed awns.   After this is done, the pasture becomes very healthy, and the herbage is nutritious.   The drought-resisting qualities of this grass are something remarkable, and often during very dry seasons it has proved a good stand-by for stock.   The panicles of this grass are often 10 inches long and they bear numerous seeds, the rigid awns of which are often more than 2 inches long.   The barb-pointed seeds of this grass are very injurious to sheep, often causing the death of numbers, by first becoming attached to the wool, then working through the skin, and often penetrating the vitals.   The seeds ripen in November, December, and January.

STIPA TUCKERII, F.v.M.

" Brown plumed spear-grass."

A slender perennial species, with rather long branching stems, and only found in scrubby country in the interior.   It is not such a strong growing grass as *Stipa elegantissima,* but it grows under somewhat similar circumstances.   The elegant plumose panicles of this grass are of a rich brown colour. This species is well worthy of introduction into gardens.   It is occasionally eaten by stock, and its seed awns are not injurious to sheep or cattle.   The seeds ripen in October and November.

TETRARRHENA JUNCEA, R. Br.

" Scrambling grass."

A perennial grass, with long slender branching stems, often scrambling over bushes to the height of 8 or 12 feet.   It is a very rare species in New South Wales.   I found it at Mossman's Bay a few years ago, which was then thought to be its only habitat in the Colony, but quite recently Dr. Woolls informs me it has been found at the National Park.   Since the foregoing was put into type I have also found it at the last mentioned place. This grass is of no value as a pasture plant; but it is of interest to the botanist, from the fact that each floret has four stamens.   Its seeds ripen during the autumn months.

TRIODIA IRRITANS, R. Br.   ˙

" Porcupine grass."

A perennial species, with long rigid round sharp-pointed leaves, and only found in the arid interior, and fortunately not very plentiful anywhere in this Colony.   This grass grows between 3 and 4 feet high, and generally on poor sandy soil.   It is one of the troublesome prickly grasses of the desert.   In letters I have received, with specimens of this grass sent me for identification, it was stated " that it has not increased for twenty years," so that it is not a very formidable enemy to vanquish.   This grass does not produce much seed.

TRIODIA MITCHELLII, Benth.

"Porcupine grass."

A perennial species, generally growing 4 to 5 feet high, which is only found in the arid interior. Like the last species, it has long rigid round sharp-pointed leaves, which are troublesome and often dangerous to man and beast. It is of no value whatever for forage, although the young growths that are made after a bush-fire are occasionally eaten by stock. When in flower, this is really a rather handsome looking grass, and any one seeing it for the first time in flower, is very much disappointed on a closer acquaintance with it. This grass produces a fair amount of seed, which ripens in the autumn months.

TRIRAPHIS MICRODON, Benth.

An erect glabrous grass of 2 or 3 feet high, which is generally found on the Blue Mountains, but it does not appear to be plentiful anywhere. It is of little value, however, for forage, as its herbage is both harsh and scanty. This grass produces a fair amount of seed, which ripens during the autumn months.

TRIRAPHIS MOLLIS, R. Br.

"Purple-headed grass."

A slender perennial species, growing from 2 to 3 feet high, which is generally found all over the interior, and in some situations it is fairly plentiful. It is a capital drought-resisting species, and during the early summer months yields a fair amount of good herbage which sheep are fond of. When the plant becomes old, however, the stems get very hard and wiry; then stock seldom or never touch it. This grass produces a rather dense panicle, often 10 inches long and of a purplish colour, which gives the plant a very ornamental appearance. It might be introduced into gardens, where it would be very effective. The seed ripens in November and December. There is a variety of this grass (var. *humilis*), which rarely exceeds 6 inches in height and has a panicle of only 2 or 3 inches. It is a somewhat uncommon grass, generally growing on sandy soil, and not of much value for forage. This delicate little grass might be introduced into gardens, where its charming purple panicles would be sure to be admired. This variety seeds in October and November.

TRISETUM SUBSPICATUM, Beauv.

"Spiked-oat grass."

A tufted perennial species, varying in height from 6 inches to above 2 feet, which is only found in the southern portions of the Colony, principally on the Muniong and other mountains. In some situations it is fairly plentiful, and is regarded as a good forage grass for sheep. It is well worthy of dissemination in the very cold parts of the Colony, for very few grasses can live where this one will flourish. It produces a fair amount of seed, which ripens in February and March. Hooker says of it, in his Antarctic Flora— "Few grasses have so wide a range as this, nor am I acquainted with any other Arctic species which is equally an inhabitant of the opposite polar regions. In Europe it is found at a very great elevation on the Alps and Pyrenees, as also in Lapland. In Asia it frequents the Altai Range, the northern parts of Siberia and Kamschatka, from whence it crosses to Kotzebue's Sound, and is apparently more generally distributed through Arctic America (than in the Old World), from the utmost limits of polar vegetation in Melville Island, throughout Greenland, and the Arctic Islands, the Arctic Sea Coast, Labrador, Canada, and the Rocky Mountains."

ZOYSIA PUNGENS, Willd.

" Coast couch grass."

A perennial species, with creeping underground stems, often to a great extent in loose sand, from which spring erect stems, rarely above 6 inches high. It is purely a coastal grass, and is never found except on the littoral sands, or in or near salt marshes. In some places it forms a compact turf, and affords a large amount of herbage, which stock are particularly fond of. Like most littoral grasses, its stems and leaves contain a considerable amount of soda, which is invaluable to the health of stock. I have often recommended this grass for planting on the littoral drift sands, and much good would have resulted had this been done at Wollongong and at Newcastle. We have a good illustration of the value of this grass for binding the littoral sands at the " spit," between North Sydney and Manly. Nearly the whole of that great sandbank has been quietly but efficiently bound together by the roots of this useful little grass, as any one may easily be convinced by taking a spade and digging down into the sand as I did, where they will see a perfect net-work of roots for several feet down. This grass should be widely distributed in the coastal districts where it is not already growing. It is easily propagated by division of its roots.

The following exotic grasses have become naturalised in various portions of the Colony :—

AIRA PRÆCOX, Linn.

" Early hair grass."

An annual species, growing from 6 to 8 inches high, which is found only in the southern portions of the Colony, but nowhere very plentiful. This grass makes most of its growth during the winter and early spring months, but its foliage soon withers on the approach of hot weather. It is not of any value for forage. Its purplish or pale-coloured flowers are borne on slender panicles, which gives the grass a rather ornamental appearance. The seeds ripen in September.

ANTHOXANTHUM ODORATUM, Linn.

" Sweet-scented vernal grass."

A perennial species, growing from 1 foot to 2 feet high, which has become fairly well established in some pastures in the southern portions of the Colony. Its presence in pastures may easily be detected by its pleasant odour. This is due to a fragrant resinous principle, called *coumarin*. It is not considered a first-class grass for forage, having a less quantity of saccharine matter and more mucilage than some other kinds in its composition. Its presence, however, in pastures is considered an advantage, especially where grasses are grown for hay, as it imparts a pleasant odour to the crop, which enhances its value. I have known of instances where cattle did eat damaged hay when flavoured with the sweet-scented vernal grass, or fenugreek (*Trigonella fœnum græcum*). Our native *hierochloas* (sweet-scented holy grasses) would answer the same purpose, as they contain the same fragrant principle (*coumarin*). The sweet-scented vernal grass does not produce a great amount of seed. What there is, however, ripens in September and October.

Mr. Martin J. Sutton gives the following analysis of this grass :—

	Grass in natural state.	Dried at 212° Fahr.
Water..	61·84	......
*Soluble albuminoids	·69	1·81
†Insoluble albuminoids..	1·31	8·44
Digestible fibre..	14·43	37·81
Woody fibre	14·56	38·15
‡Soluble mineral matter	1·76	4·61
‖Insoluble mineral matter	·83	2·18
Chlorophyll, soluble carbo-hydrates, &c.	4·58	12·00
	100·00	100·00
*Containing nitrogen	·11	·29
†Containing nitrogen	·21	·55
Albuminoid nitrogen	·32	·84
Non-albuminoid nitrogen	·20	·55
Total nitrogen..................	·52	1·39
‡Containing silica..	·33	·99
‖Containing silica..	·44	1·15

AVENA FATUA, Linn.
" Wild oats."

An annual species, growing about 3 feet high, and generally found about stockyards both in the coastal districts and far into the interior. During the winter and early spring months it yields a lot of succulent herbage, which cattle browse upon. When the hot weather sets in, however, the stems become very hard; then cattle seldom or never touch it. In England several years ago experiments were carried out with this grass with a view of improving the grain, and after several years of careful selection and good cultivation, it produced a grain equal to the present cultivated oat. In fact, many botanists are of opinion that the cultivated oat is a domesticated variety of the wild species. The wild oat produces a fair amount of seed when left undisturbed for a time, and it ripens in October and November.

BRIZA MAXIMA, Linn.
" Large quaking grass.

An annual species, rarely exceeding a foot in height, and generally found in the coastal districts. It makes its growth during the early spring months, but it is of no value from an agricultural point of view, as stock seldom or never touch it. It is a very ornamental grass, however, when in seed, and it may often be seen cultivated in gardens. The imbricated flowers are arranged in panicles, and often are gathered for decorative purposes. The seeds ripen in September and October.

BRIZA MINOR, Linn.
" Little quaking grass."

An annual species, rarely exceeding 9 inches in height, and generally found all over the Colony, especially near cultivation, and in some situations it is fairly plentiful. It is only found in the spring of the year, for on the advent of hot weather it rapidly disappears. This grass is cultivated in gardens, and its ornamental panicles often are gathered for decorative purposes, though not to the extent of the last-named species. The seeds ripen in September and October.

E

BROMUS MOLLIS, Linn.
" Soft brome grass."

An annual species, growing from 1 foot to 2 feet high, which is generally found about old stock-yards in various portions of the Colony, but nowhere very plentiful. It is a purely winter and early spring grass, for on the advent of hot weather it rapidly dies away. It is not of any agricultural value, however, as stock do not care about its hairy, soft leaves, and will not eat it when other grasses are plentiful. The seeds ripen in October and November.

BROMUS STERILIS, Linn.
" Barren brome grass."

An annual species, growing from 2 to 3 feet high, which is found only in the coastal districts, but as far as I have been able to find out it is not very plentiful. It makes most of its growth during the winter and early spring months, but dies on the approach of hot weather. Although cattle will graze upon its soft, downy herbage, still it cannot be regarded as a good pasture grass, and they will not touch it when other herbage is plentiful. This grass does not perfect much seed. What little there is, however, ripens in October and November.

CERATOCHLOA UNIOLOÏDES, D'C.
" Prairie grass."

An annual species, growing from 2 to 3 feet high, which is generally found in the coastal districts, and in some places it is very plentiful; in fact, it is this grass which keeps some of the parks of Sydney beautifully green throughout the winter months. I have no hesitation in saying that this is the best annual grass ever introduced into Australia. From the sweetness of its taste and the greediness with which it is eaten by all herbivora, there can be little doubt that it is a very nutritious grass. It starts into growth after the first autumn rains, and continues throughout the winter and early spring months, until November, when it gradually dies away. During its growing period, however, it yields a phenomenal amount of rich succulent herbage, and if cut before the flower-stems appear, it can be made into excellent hay. This grass will stand close-feeding, and if a fair-sized paddock were judiciously penned off, a good flock of sheep could be kept in splendid condition on it for about seven months in the year. It is a prolific seed-bearing grass, and the seeds ripen in October and November. Baron F. von. Mueller and L. Rummel give the following chemical analysis of the spring growth of this grass:—Albumen, 2·80; gluten, 3·80; starch, 3·30; gum, 1·70; sugar, 2·30 per cent.

DACTYLIS GLOMERATA, Linn.
" Cock's-foot grass or orchard grass."

A perennial species, growing from 1 foot to 3 feet high, which is generally found in the colder parts of the Colony. In the winter and early spring months it yields a tender herbage, which stock of all descriptions are fond of. To render it of any value in permanent pastures, however, it should be kept well eaten down, for, if allowed to grow any length of time unmolested, it forms large tufts of coarse herbage, which stock will not relish whilst other grasses are plentiful. It is not a good grass to sow in mixtures with others, as, from its coarse habit, it generally crowds out the more delicate ones, and it should never be sown on poor or dry soils. Under these circumstances, its herbage is always stunted and wiry, and stock seldom touch it. The only places where this grass does well in this Colony is on damp, rich, strong soils, in the coldest parts of the Colony. It produces a fair amount of seed if left unmolested for a time, and it ripens during October and November. Baron von Mueller and L. Rummel give the following chemical analysis made on the late spring growth of this grass :—Albumen, 1·87 ; gluten, 7·11 ; starch, 1·05 ; gum, 4·47 ; sugar, 3·19 per cent.

Mr. Martin J. Sutton gives the following chemical analysis of this grass :—

	Grass in natural state.	Dried at 212° Fahr.
Water	60·74	....
*Soluble albuminoids	·25	·62
†Insoluble albuminoids	1·50	3·81
Digestible fibre	11·30	28·78
Woody fibre	16·24	41·36
‡Soluble mineral matter	2·04	5·19
‖Insoluble mineral matter	·91	2·32
Chlorophyll soluble carbo-hydrates, &c.	7·02	17·92
	100·00	100·00
*Containing nitrogen	·04	·10
†Containing nitrogen	·24	·61
Albuminoid nitrogen	·28	·71
Non-albuminoid nitrogen	·18	·46
Total nitrogen	·46	1·17
‡Containing silica	·35	·89
‖Containing silica	·51	1·29

FESTUCA RIGIDA, Mert. et Koch.

"Rigid fescue."

A small, rigid, tufted, annual grass, rarely exceeding 6 inches in height, and is extremely rare in this Colony, but quite common in the southern Colonies. The only specimens I have seen here are one which I collected in Messrs. Shepherd's Nursery, and the other which Mr. Wooff, of Victoria Park, sent me for identification. If this grass gets well established here, however, it will prove troublesome in cultivated ground. It produces a fair amount of seed, which ripens in October and November.

HOLCUS LANATUS, Linn.

"Yorkshire fog or meadow soft grass."

A perennial species, growing about 2 feet high, which is generally found in the colder parts of the Colony, and in some places fairly plentiful. It is not a good pasture grass, although it makes considerable growth in early spring, which herbage however, is rather disliked by stock of all kinds, and whilst other grasses are plentiful they will not eat it. This grass is easily recognised in pastures by its pale, soft appearance. It produces an abundance of seed, which, when ripe, is easily disseminated by winds. The seeds ripen in November and December. Baron von Mueller and L. Rummel, give the following analysis, made on the spring growth of this grass :—Albumen, 3·20 ; gluten 4·11 ; starch, 0·72 ; gum, 3·08 ; sugar, 4·56 per cent.

HORDEUM MURINUM, Linn.
" Barley grass."

An annual species, growing from 12 to 18 inches high, which is generally found all over the Colony, and about old stock-yards it is very plentiful. During the winter and early spring months it yields a fair amount of succulent herbage, which stock of all kinds will graze upon; but in the early summer months it gets very hard and wiry; then stock will not touch it. It is not a desirable grass to have in the pastures, for the seeds, with their adherent awns, which often are an inch long, are troublesome to the salivary glands of sheep and other small herbivora. It is a prolific seed-bearing grass, and the seeds ripen in October and November.

HORDEUM NODOSUM, Linn.
" Meadow barley."

A perennial species, growing about 2 feet high, and only found at a few places in the southern portion of the Colony. Although a taller grass than the last-named species, it is not so coarse, neither are its seed awns so long; and taken altogether, it is a much superior pasture grass to the last-named species. When left unmolested for a time, it produces a fair amount of seed, which ripens in November and December.

KOELERIA CRISTATA, Pers.
" Crested-hair grass."

A perennial species, growing from 1 foot to 3 feet high, which is found only very sparingly in New South Wales. It is not a valuable pasture grass, as it contains little nutriment; but as it will grow on dry soils and sustain itself during very dry weather, it might prove of value when other herbage is scarce. It produces a fair amount of seed, which ripens in November and December.

KOELERIA PHLEOIDES, Pers.

An annual grass, growing from 6 inches to a foot high, which is only found very sparingly in this Colony. It is of no value from an agricultural point of view, as its herbage soon withers off on the approach of hot weather. It does not produce much seed: what little there is, however, ripens in the early summer months.

LOLIUM PERENNE, Linn.
" Perennial rye grass."

This is probably the best known of all agricultural grasses. It has been stated on good authority that it was cultivated in the 17th century, and has been cultivated more or less ever since that time. As might be supposed, there are now quite a number of varieties—by some authorities computed at over fifty. The Italian rye-grass (*Lolium italicum* of some botanists) is one of its many varieties. As a winter and early spring grass, it is a valuable addition to the pastures in the southern and colder districts of the Colony, but it will not grow satisfactorily west of the Dividing Range, neither will it thrive on poor dry soils. An alluvial, moist, rich, strong land suits it best, and under such conditions it will yield a bulk of valuable herbage, of which milch cows are particularly fond. It is said, however, that sheep do not always thrive well on it, and they sometimes are subject to fits similar to those produced by eating the darnal grass when kept too long on pastures where the rye grass predominates. This, however, may be caused by a diseased or ergotised state of the rye or other grasses, which I have often pointed out is a source of very great danger in our pastures. Rye-grass will stand almost any amount of irrigation during our hot summers, and unless this is done it will cease to grow and get very brown about midsummer. It also makes capital hay if cut when in flower. The rye-grass produces an abundance of seed, which ripens in October and November.

Baron von Müeller and L. Rummel give the following chemical analysis, made on the spring growth of this grass :—Albumen, 3·36 ; gluten, 4·88 ; starch, 0·51 ; gum, 1·80 ; sugar, 1·80 per cent.

Mr. Martin J. Sutton gives the following analysis of this grass :—

	Grass in a natural state.	Dried at 212° Fahr.
Water ..	62·01	......
*Soluble albuminoids	·38	1·00
†Insoluble albuminoids	2·06	5·38
Digestive fibre	7·98	21·01
Woody fibre	17·71	46·62
‡Soluble mineral matter	2·90	7·64
‖Insoluble mineral matter	·78	2·05
Chlorophyll, soluble carbo-hydrates, &c.	6·18	16·30
	100·00	100·00
*Containing nitrogen	·06	·16
†Containing nitrogen	·33	·86
Albuminoid nitrogen	·30	1·02
Non-albuminoid nitrogen	·38	1·00
Total nitrogen	·77	2·02
‡Containing silica	·05	·13
‖Containing silica	·32	·84

LOLIUM TEMULENTUM, Linn.
" Darnal grass."

An annual species, which is gradually spreading in the coastal districts in various parts of the Colony, much to the concern of some pastoralists. This is a deleterious grass which sometimes prevails to a dangerous extent in pastures. It is said to produce poisonous effects on the system, such as headache, drowsiness, vertigo, &c., &c. I have had several specimens sent for identification from various parts of the Colony, and I have collected it in the vicinity of Sydney. When left unmolested for a time it produces a great amount of seed, which ripens in November and December.

POA ANNUA, Linn.
" Goose grass."

An annual species, of tufted habit, sometimes only 2 or 3 inches, at other times nearly 12 inches high, which is generally found all over the Colony, with the exception of the arid interior. It generally springs up after the autumn rains, and continues to grow throughout the winter and early spring months, but dies off on the approach of hot weather. It is of no value from an agricultural point of view, but it imparts a pleasing green to many, what would otherwise be, dreary spots in the winter and early spring months, This grass can be recommended for covering bare patches under the shade of trees. It is a prolific seed-bearing grass. The seeds ripen in September and October.

POA GLAUCA, E.B.
" Glaucous poa."

As far as is at present known, this grass has not made much headway in the country, so that very little is known of its value as a pasture grass from an Australian point of view.

POA PRATENSIS, Linn.
" English meadow grass, or Kentucky blue grass."

A perennial species, which is spreading very rapidly in the coastal districts. It may generally be found on light dry soils, where its underground stems can easily ramify, and they soon form a perfect mat. It is a capital grass to withstand dry weather, and its dark green foliage may be seen when the surrounding grasses are dried off. I have used this grass for making lawns and binding embankments, and I can highly recommend it for such purposes; but I cannot advise it to be sown in pastures that are occasionally brought under cultivation, for its underground roots are most difficult to exterminate, and would very soon smother a tender crop. Although it makes splendid verges to gravel walks, or flower borders, still it cannot be recommended, for the underground stems are difficult to keep in order. This grass is a rather shy seed bearer. It produces its flower panicles only once during the season. The seeds ripen in November and December. It is easily propagated, however, by the division of its roots.

Mr. Martin J. Sutton gives the following analysis of this grass :—

	Grass in a natural state.	Dried at 212° Fahr.
Water...	65·81	......
Soluble albuminoids	......	......
*Insoluble albuminoids..............................	1·81	5·31
Digestible fibre......................................	9·29	27·17
Woody fibre	15·24	44·57
†Soluble mineral matter	1·11	3·24
‡Insoluble mineral matter	1·42	4·13
Chlorophyll, soluble carbo-hydrates, &c. ...	5·32	15·58
	100·00	100·00
*Containing nitrogen	·29	·85
Non-albuminoids nitrogen........................	·15	·44
Total nitrogen................	·44	12·9
†Containing silica	·40	1·17
‡Containing silica	1·13	3·29

POLYPOGON MONSPELIENSIS, Desf.
" Beard grass."

An annual species, growing from 1 foot to 2 feet high, which is generally found in the coastal districts, but nowhere very plentiful. It is of no agricultural value, however, and would be likely to become a pest in cultivated ground if allowed head-way. It produces a fair amount of seed, which ripens in October and November.

PHALARIS CANARIENSIS, Linn.
" Canary grass."

An annual species, which is now widely distributed throughout the Colony, and in some places during the early summer months it may be seen in fair quantities. This grass produces the canary seed of commerce, and

I recently have seen some very fine samples that were grown in the Colony. This grass would pay to cultivate to supply our local market, but not for export. It is of comparatively easy culture, and only occupies the ground for a few months of the year. The soil best suited to its growth is a good open sandy loam, not too rich, otherwise it will make a quantity of leaf and stem at the expense of the seed. It is not a good pasture grass, for it takes such a slight hold of the ground that cattle grazing upon it often pull it up by the roots. The following chemical analysis was made on the November growth of this grass by Baron von Mueller and L. Rummel:—Albumen, 1·59; gluten, 6·14; starch, 1·03; gum, 6·64; sugar, 2·86 per cent. Another analysis made by the same gentlemen in the same month gave :—Albumen, 1·06; gluten, 5·64; starch, 0·98; gum, 3·22; sugar 4·20 per cent.

PHALARIS MINOR, Retz.
" Lesser canary grass."

An annual species, widely distributed in the coastal districts, but not of any value from an agricultural point of view. It produces a fair amount of seed, which is like the smallest grains of canary-grass, and might be used for a somewhat similar purpose. The seeds ripen in November and December.

STENOTAPHRUM AMERICANUM, Schr.
Here called the " buffalo-grass."

A rather coarse perennial species, creeping and rooting at the base, and rarely exceeding 1 foot in height. It attains its greatest perfection in the coastal districts when grown on loose soils. It will not bear much frost, neither will it grow well in the arid interior. It is admirably adapted for making lawns, garden-edgings, binding river banks against the fury of flood-waters, and it is not destroyed by being submerged for a few days. Once it gets well established it is difficult to eradicate. It was this grass that my friend, Mr. J. C. Bell, reared with so much advantage on the Island of Ascension, where it was thought hardly anything would grow. There is much diversity of opinion as regards its value for forage, some asserting that it makes splendid feed, whilst others argue the very opposite. From personal experience, which has extended over a number of years, I may say that pasture animals will not eat it when other herbage is plentiful. Baron von Mueller and L. Rummel give the following analysis, made on the late spring growth of this grass :—Water, 80·25 ; albumen, 0·50; gluten, 5·44; starch, 0·08 ; gum, 1·60 ; sugar, 1·60 ; fibre, 10·53 per cent.

Index to Census of Indigenous Grasses.

Lepturus cylindricus. *Trin.*
,,   incurvatus. *Trin.*
Microlæna stipoides. *R. Br.*
Neurachne alopecuroides. *R. Br.*
,,   mitchelliana. *Nees.*
,,   munroi. *F. v. M.*
Oplismenus compositus. *Beauv.*
,,   setarius. *R. et S.*
Panicum adspersum. *Trin.*
,,   baileyi. *Benth.*
,,   bicolor. *R. Br.*
,,   cœnicolum. *F. v. M.*
,,   crus-galli. *Linn.*
,,   decompositum. *R. Br.*
,,   distachyum. *Linn.*
,,   divaricatissimum. *R. Br.*
,,   effusum. *R. Br.*
,,   flavidum. *Retz.*
,,   foliosum. *R. Br.*
,,   gracile. *R. Br.*
,,   helopus. *Trin.*
,,   indicum. *Linn.*
,,   leucophœum. *H. B. et K.*
,,   macractinum. *Benth.*
,,   marginatum. *R. Br.*
,,   melananthum. *F. v. M.*
,,   mitchellii. *Benth.*
,,   obseptum. *Trin.*
,,   parviflorum. *R. Br.*
,,   prolutum. *F. v. M.*
,,   pygmæum. *R. Br.*
,,   repens. *Linn.*
,,   reversum. *F. v. M.*
,,   sanguinale. *Linn.*
,,   semialatum. *R. Br.*
,,   tenuissimum. *Benth.*
,,   trachyrachis. *Benth.*
,,   uncinulatum. *R. Br.*
Pappophorum avenaceum. *Lindl.*
,,   nigricans. *R. Br.*
Paspalum brevifolium. *Flügge.*
,,   distichum. *Linn.*

Paspalum minutiflorum. *Steud.*
,,   scrobiculatum. *Linn.*
Pennisetum compressum. *R. Br.*
Pentapogon billardierii. *R. Br.*
Perotis rara. *R. Br.*
Phragmites communis. *Trin.*
Poa cæspitosa. *Forst.*
,, lepida. *F. v. M.*
,, nodosa. *Nees.*
Pollinia fulva. *Benth.*
Potamophila parviflora. *R. Br.*
Schedonorus hookerianus. *Benth.*
,,   littoralis. *Beauv.*
Setaria glauca. *Beauv.*
,,   macrostachya. *H. B. et K.*
Sorghum halepense. *Pers.*
,,   plumosum. *Beauv.*
Spinifex hirsutus. *Labill.*
,,   paradoxus. *Benth.*
Sporobolus actinocladus. *F. v. M.*
,,   diander. *Beauv.*
,,   indicus. *R. Br.*
,,   lindleyi. *Benth.*
,,   pulchellus. *R. Br.*
,,   virginicus. *Kunth.*
Stipa aristiglumis. *F. v. M.*
,,   elegantissima. *Labill.*
,,   flavescens. *Labill.*
,,   micrantha. *Cav.*
,,   pubescens. *R. Br.*
,,   scabra. *Lindl.*
,,   semibarbata. *R. Br.*
,,   setacea. *R. Br.*
,,   tuckerii. *F. v. M.*
Tetrarrhena juncea. *R. Br.*
Triodia irritans. *R. Br.*
,,   mitchellii. *Benth.*
Triaphis microdon. *Benth.*
,,   mollis. *R. Br.*
Trisetum subspicatum. *Beauv.*
Zoysia pungens. *Willd.*

Index to List of Exotic Grasses which have become naturalised in various parts of the Colony.

Aira præcox. *Linn.*
Anthoxanthum odoratum. *Linn.*
Avena fatua. *Linn.*
Briza maxima. *Linn.*
,, minor. *Linn.*
Bromus mollis. *Linn.*
,, sterilis. *Linn.*
Ceratochloa unioloides. *D'C.*
Dactylis glomerata. *Linn.*
Festuca rigida. *Mert. et Koch.*
Holcus lanatus. *Linn.*
Hordeum murinum. *Linn.*

Hordeum nodosum. *Linn.*
Koeleria cristata. *Pers.*
,, phleoides. *Pers.*
Lolium perenne. *Linn.*
,, temulentum. *Linn.*
Poa annua. *Linn.*
,, glauca. *E. B.*
,, pratensis. *Willd.*
Polypogon monspeliensis. *Desf.*
Phalaris canariensis. *Linn.*
,, minor. *Retz.*
Stenotaphrum americanum. *Schr.*

Sydney : Charles Potter, Government Printer.—1890.